Aerodynamics Principles for Air Transport Pilots

Aerodynamics Principles for Air Transport Pilots

Rose G. Davies

CRC Press is an imprint of the
Taylor & Francis Group, an **informa** business

First edition published 2020

by CRC Press
6000 Broken Sound Parkway NW, Suite 300, Boca Raton, FL 33487-2742

and by CRC Press
2 Park Square, Milton Park, Abingdon, Oxon, OX14 4RN

International Standard Book Number-13: 978-0-367-18854-2 (Hardback)
International Standard Book Number-13: 978-0-429-26115-2 (eBook)

Library of Congress Cataloging-in-Publication Data

[Library of Congress Control Number: 2020932610]

Typeset in Palatino LT Std
by Deanta Global Publishing, Services, Chennai, India

Contents

Preface

It is extremely important for students aspiring to be air transport pilots to clearly understand aerodynamic principles. This is best achieved by commencing with fundamental principles and progressing into the necessary use of the theories in physics and mathematical analysis needed to acquire the appropriate level of knowledge. This is, in particular, beneficial for tertiary pilot students.

Aerodynamics has been taught to pilots in various ways, depending on the category of the licensing examination. The aerodynamics courses offered to Bachelor of Aviation (BAv) students in the Air Transport Pilot Program at Massey University in New Zealand are structured such that the theory is linked to applied aerodynamics. Such a course structure equips the students with the ability to relate theory to practical flight and enables them to better analyse real-life situations. These courses are not just designed for BAv students; they can also provide professional air transport pilots with guidance, based on theory, to understand abnormal issues encountered in practice. The content of this book has been developed from courses delivered at Massey University, the scope of the book being to provide a bridge between the academic content of classical aerodynamics and the practical phenomena encountered during aircraft operations. The learning material in this book is presented in three major sections, namely incompressible airflow, compressible airflow, and supersonic waves. This book covers the requirements for aerodynamics up to and beyond the level required for an Air Transport Pilot Licence (ATPL).

Chapter 1 is revision of the calculus required for this book. Chapters 2, 3, 4, and 5 deal with incompressible fluid dynamics, basic principles in thermodynamics, and viscous flow. The aerodynamic aspects of subsonic flight, namely lift, drag, and stall, are explained using these principles, specifically the relationship between angle of attack (AoA), lift coefficient, and airspeed. Subsequently, there is discussion on the aerodynamic functionality of features of aerofoils and other devices on the wings, followed by the effects of these on lift coefficient, drag coefficient, stall prevention, and aircraft stability.

The aerodynamics principles for compressible airflow and their applications are introduced and discussed in Chapters 6, 7, 8, and 9. The changes of airspeed and related air pressure, as well as the formation of shockwaves are the main focus of this section. Applications of compressible flow aerodynamics, its effects on mechanics of flight of high-subsonic and transonic aircraft, transonic aerofoils and the flight control issues of transonic flights are covered. Students are encouraged to develop methods to investigate aerodynamic forces on any aerofoil to explain the performance of transonic aircraft, and, in turn, to demonstrate their understanding of problems that are unexpected, and complicated events that may occur during flight.

In the final section of this book (Chapters 10 and 11), oblique shockwave and expansion waves in a supersonic airflow are analysed in relation to the conditions at their formation, their characteristics, and the changes of air property aft of the waves. The characteristics of the shockwaves and expansion waves around supersonic aerofoils is analysed using the features of these supersonic waves. The intention is to provide students with a grasp of the aerodynamic principles of lift production, so that they can comprehend the fundamental differences in the principles of aerodynamic forces for different aircraft from the low subsonic to the supersonic range.

It will be advantageous to students if they have some knowledge of fundamental physics and basic calculus, and preliminary flight experience before taking the aerodynamics course based on the content of this book.

Acknowledgment

I owe a debt of gratitude to my students, who were the "guinea pigs" in my journey to develop the concept of this book. When I produced the content of each chapter, I tested it on them. They faithfully gave me their feedback. I appreciate greatly their participation. I would like to express my gratitude to my colleagues at the School of Aviation, Massey University, who have been very supportive, creating a work environment where I could concentrate on teaching and writing, the flight instructors, and technicians and engineers at our hangar, who have been tirelessly answering my questions; I have learned a great deal from them.

I thank Dr Tony Howes at the University of Queensland, Australia, who volunteered to be my proof-reader. His rich knowledge in fluid dynamics, physics, and engineering kept my thoughts on the right track. I thank Mr Mohammad Seraj, who assisted me with creating the illustrations required in this book. I appreciate his artistic ability and professionalism.

Lastly and importantly from the bottom of my heart, I thank Professor Clive Davies, my husband, who was my first reader, for his continual encouragement and unfailing support and love.

Author

Rose G. Davies works at the School of Aviation, Massey University, New Zealand. She developed, and coordinates the current aerodynamics courses for the BAv degree in the Massey University Air Transport Program. She has teaching experience in aero-science and aircraft systems, physics and mathematics, and the foundation courses for various degrees. She has a Bachelor's degree in mechanical engineering, majoring in internal combustion engine design, a Master's degree in engineering-thermophysics–combustion, and a PhD in applied mathematics. Before starting her university teaching career, Rose had some 20 years' research experience in mathematical modeling and fluid dynamics, combustion, and remote sensing of engine emissions. She is a member of ASME, AIAA, ANZIAM, and the Royal Society New Zealand, and a member of the Editorial Board of the Journal of Aviation/Aerospace Education and Research (JAAER).

Notation

a	Acceleration, speed of sound
A	Area
AC	Aerodynamic center
b	Wing span
c	Chord
C_L	Lift coefficient
C_D	Drag coefficient
C_p	Pressure coefficient
C_p	Specific heat while pressure is constant
C_V	Specific heat while volume is constant
CP	Center of pressure
D	Drag
E	Energy
h	Height
L	Lift
M	Mach number; Molar mass
m	Mass
p	Pressure
Q	Heat
R	Universal gas constant
Re	Reynolds number
R_M	Gas constant
S	Area of wing
t	Time
T	Temperature
V	Air flow speed, velocity
V	Volume
W	Weight; Work
x	Displacement in horizontal or along stream line
y	Displacement in vertical or perpendicular to the surface of aerofoil
α	Angle of attack
γ	Adiabatic index
δ	Thickness of boundary layer; deflection angle
λ	Dynamic viscosity of fluid
μ	Dynamic viscosity; *Mach* angle
ν	Kinematic viscosity
ρ	Density
θ	Deflection angle
β	Shock angle
ϕ	Diameter

1

Calculus Revision

Calculus is a convenient tool in aerodynamics. It aids in explaining the characteristics of functions, which describe the airflow fields. It is assumed that students have gained the skills to differentiate basic functions from previous studies. Readers should have learnt the method to differentiate basic functions – for example, polynomials, logarithmic, trigonometric, and exponential functions. This chapter concentrates on explaining some applications of calculus, including the meanings of derivatives of a function in real life, an analysis of the changes of a function with two or more variables, and simple concepts of integration. This chapter will use plain language as much as possible so that readers can understand the mathematic expressions when the concepts discussed are used in explaining aerodynamics principles.

Differentiation

In calculus, you have learnt differentiation. It is assumed that readers are able to find the derivatives of a function. The following example shows how to revise the meanings of derivatives in physics.

Assume a displacement (distance) y in m, of an object is a function of time t in s, and the function is continuous, as shown in Figure 1.1. The function is:

$$y = y(t) = t^3 - 2t^2 - 2t + 1 \tag{1.1}$$

where t is a variable and y is a function of t. Figure 1.1 illustrates this function, $t \geq 0$.

To differentiate y with respect to t produces the derivative $y' = \dfrac{dy}{dt}$. The derivative y' indicates the change rate of the original function y. The change rate y' of the displacement function y is the velocity (speed), v, of the object. y' is obtained by using $(t^n)' = nt^{n-1}$:

$$v = y' = \frac{dy}{dt} = 3t^2 - 4t - 2 \tag{1.2}$$

Figure 1.2 shows the velocity function v with respect to time t.

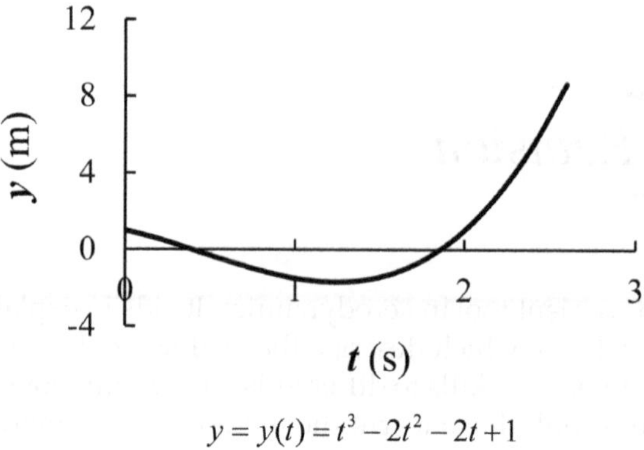

FIGURE 1.1
A displacement function $y(t)$.

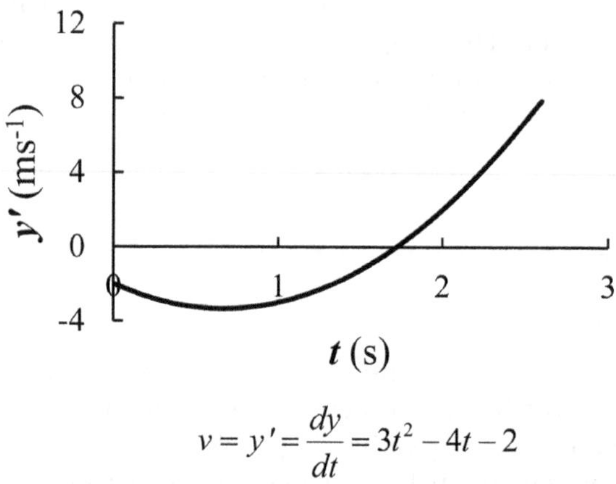

FIGURE 1.2
The derivative function $y'(t)$ of $y(t)$.

We differentiate y' to get $y'' = \dfrac{d^2y}{dt^2}$. y'' is the change of y', i.e. the change of velocity of the object. The change of velocity is the acceleration, a, of the object. The derivative of y' can be obtained by the same process as shown in equation (1.2):

$$a = y'' = \frac{dv}{dt} = \frac{d^2y}{dt^2} = 6t - 4 \tag{1.3}$$

Figure 1.3 shows acceleration function **a** with respect to time t.

The acceleration, a, is the first-order derivative of the velocity, v, and is the second-order derivative of the displacement, y. The displacement y is a cubic function (polynomial) of time t, and the velocity, v, is a quadratic function of t, and the acceleration, a, is a linear function of t.

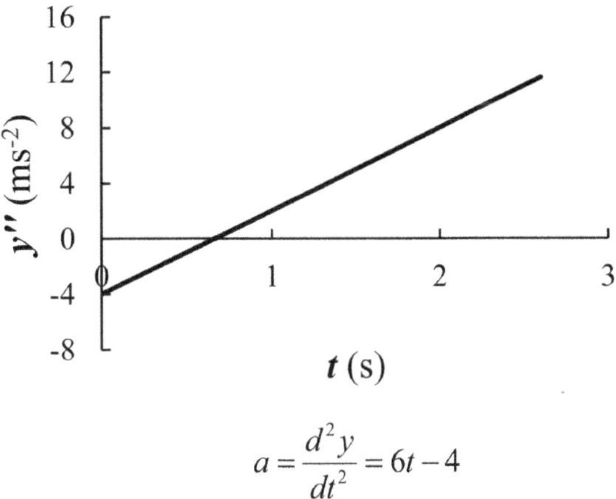

$$a = \frac{d^2y}{dt^2} = 6t - 4$$

FIGURE 1.3
The second-order derivative function $y''(t)$ of $y(t)$.

In summary, the derivative of a function expresses the rate of change of the function with its variable. The change of displacement with time is velocity; the change of mass with time is mass flowrate; and the change of a volume with time is the volumetric flowrate.

If C is a constant with respect to a variable, for example, t, the derivative of $C' = \frac{dC}{dt} = 0$. This means that there is no change. For example, when the acceleration of an object is 0, i.e. the derivative of its velocity is 0, it means that the velocity is constant.

Function analysis (derivatives)

One of the applications of differentiation-calculus is to analyse a function. According to the principle of differentiation discussed in *Calculus: Early Transcendentals* (Stewart, J., Thomson Higher Education, 2009), the derivative of a function at any point on the function is the rate of change of this function with respect to its variable. Use the example shown earlier: when the velocity of the object is positive at $t = 2$ s (see Figure 1.2), the displacement of the object is increasing, as shown in Figure 1.1. When the velocity is negative at $t = 1$ s (see Figure 1.2), the displacement of the object is decreasing, as shown in Figure 1.1. If the acceleration, i.e. the derivative of velocity, of an object is positive, > 0, the velocity of the object is increasing; if the acceleration is negative, < 0, i.e. decelerating, the velocity of the object will be decreasing.

In general terms: for a function of x, $f(x)$, the derivative of the function $f'(x)$ is the slope of the tangent line at x. The signs of its derivative $f'(x)$ and some tangent lines are marked along the function $f(x)$ in Figure 1.4.

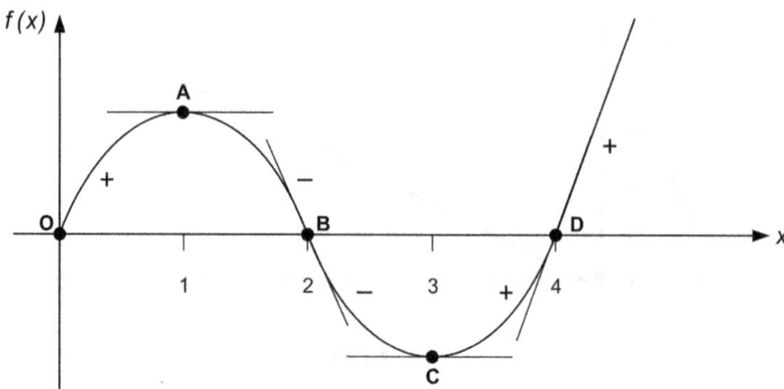

FIGURE 1.4
A function with the features of its derivative.

For this function, Figure 1.4 shows that:

when x is in the region between O and A, $0 > x > 1$, and the region between C and D, $x > 3, f'(x) > 0$, "+", and the function $f(x)$ increases with x;

when x is in the region between A and C, $1 > x > 3, f'(x) < 0$, "–", and the function $f(x)$ decreases with x.

It can also be observed in Figure 1.4 that the derivative of the function changes from "+" to "–", or changes from "–" to "+" smoothly, which means that there is a point of the derivative $f'(x) = 0$ between a positive derivative region and a negative derivative region. For example, at point A, point C, the derivative of the function is "0", i.e. the change of the function at those points is "0". Therefore, those points are ***extrema*** (which can also be called ***stagnation points***). The values of the function at the extrema are extreme values.

Extreme values

It is important in actual practice to be able to identify where or when the extreme values of a function occur, because this information might be able to optimise the advantages, or minimise the risks, or indicate some limits.

As indicated in previously, extrema occur at the derivative of a continuous function, "0".

The following shows how to identify the type of extrema of $f(x)$, a function of x:

If $f'(x) = 0$, at $x = x_o$ (e.g., Point A in Figure 1.4, $x_o = 1$.), the derivative before this point is "–", i.e. $f'(x) < 0$, and the function $f(x)$ is decreasing; the derivative after this point is "+", i.e. $f'(x) > 0$, and the function $f(x)$ is increasing, the value of the function at this point $f(x_o)$ is a local minimum; or if $f'(x) = 0$, and $f''(x) > 0$, at $x = x_o$, this $f(x_o)$ is a local minimum.

If $f'(x) = 0$, at $x = x_o$ (e.g., Point C in Figure 1.4, $x_o = 3$.), and the derivative before this point is "+", i.e. $f'(x) < 0$, and the function $f(x)$ is increasing; the derivative after this point is "–", i.e. $f'(x) > 0$, and the function $f(x)$ is decreasing, the value of the function at this point $f(x_o)$ is a local maximum; or if $f'(x) = 0$, and $f''(x) < 0$, at $x = x_o$, this $f(x_o)$ is a local maximum.

Please note that those extreme values are called local or relative maximum/minimum because the absolute maximum/minimum values can occur at other places. For some functions, the values of the functions can go to infinity. For example, in Figure 1.4, the value of the function $f(x)$ can be greater than $f(1)$ (Point A, the local maximum), if $x \gg 4$ (after point D).

Example 1.1

$y = y(t) = t^3 - t^2 - 2t + 1$. Find the extreme values of the function.

SOLUTION

Set $v = \frac{dy}{dt} = y' = 3t^2 - 2t - 2 = 0$

Use the formula to obtain quadratic roots: $t = \frac{-b \pm \sqrt{b^2 - 4ac}}{2a}$ for equation: $at^2 + bt + c = 0$:

$t = \frac{2 \pm \sqrt{(-2)^2 - 4 \times 3 \times (-2)}}{2 \times 3} = \frac{1}{3} \pm \frac{\sqrt{7}}{3}$, then $t = \frac{1}{3} + \frac{\sqrt{7}}{3} = 1.215$, and $t = \frac{1}{3} - \frac{\sqrt{7}}{3} = -0.55$.

TWO SOLUTIONS

When $t < -0.55$, for example, $t = -1$, $y'(-1) = 3 > 0$; the function is increasing.
When $t > -0.55$, for example, $t = 0$, $y'(0) = -2 < 0$; the function is decreasing.
So at $t = -0.55$, y is a local maximum.
Check $y'' = -5.3 < 0$ at $t = -0.55$, and it is confirmed that it is a local maximum.
When $t < 1.215$, for example, $t = 1$, $y'(1) = -1 < 0$; the function is decreasing.
When $t > 1.215$, for example, $t = 2$, $y'(2) = 6 > 0$; the function is increasing.
So at $t = 1.215$, y is a local minimum.
Check $y'' = 5.29 > 0$ at $t = 1.215$, and it is confirmed that it is a local minimum.

Example 1.2

Maximum area: A rectangular area created by a fence. The total length of the fence is 100 m. One side of the rectangular is x, as shown in the diagram below. Find the values of x and the area of the rectangular when the area is the optimum.

x

$50 - x$

SOLUTION

The area of the rectangular is $A = x(100/2 - x)$.
Differentiate A:

$$\frac{dA}{dx} = (50 - x) - x; \text{ and set } \frac{dA}{dx} = 50 - 2x = 0.\ x = 25 \text{ (m)}.$$

$$A'' = \frac{d^2A}{dx^2} = -2 < 0,$$

$\therefore$when $x = 25$ (m), the area of the rectangular is the relative maximum: $A_{max} = 25 \times (50 - 25) = 625$ m^2.

Derivatives of functions of multi-variables

In general, the functions discussed in aerodynamics are the functions of two or more variables. Typically, for example, velocity v of fluids is the function of location (x, y, and z) and time t: $v = v(x, y, z, t)$. Furthermore, the variable of the location of the fluid can be a function of time as well, i.e. $x(t)$, $y(t)$ and $z(t)$.

A multi-variable function can be expressed as $F = f(x, y, z)$, for example, air velocity, v, can vary in horizontal direction and vertical direction. We can take the horizontal direction as x direction and the vertical one as y direction, so $v = v(x, y)$ The air velocity may change very rapidly in vertical direction, while it remains constant in horizontal direction. This means that the derivative of v in y direction is different from that in x direction. In general, the changes of the function with different variables are different. It means that differentiating a function with respect to different variables of the function results in different functions. This differentiating process is known as ***partial differentiation***. Partial differentiation deals with one variable at a time to produce partial derivatives:

$$\frac{\partial F}{\partial x}, \frac{\partial F}{\partial y}, \text{ and } \frac{\partial F}{\partial z}$$

to describe the changes of the function in different directions if x, y, and z are the coordinates in three directions of space.

Partial differentiation

To obtain a partial derivative is to differentiate a function with respect to a variable, and the other variables are treated as constants.

For example, if $f(x, y) = x^2 + 3y$, $\frac{\partial f}{\partial x} = 2x$ and $\frac{\partial f}{\partial y} = 3$.

The same principle will be applied to high-order differentiation; for the same example as above, $\frac{\partial^2 f}{\partial x^2} = 2$ and $\frac{\partial^2 f}{\partial y^2} = 0$.

Cross second-order partial derivatives can also be determined: $\frac{\partial^2 f}{\partial x \partial y} = \frac{\partial}{\partial x}\left(\frac{\partial f}{\partial y}\right)$ and $\frac{\partial^2 f}{\partial y \partial x} = \frac{\partial}{\partial y}\left(\frac{\partial f}{\partial x}\right)$, and $\frac{\partial^2 f}{\partial x \partial y} = \frac{\partial^2 f}{\partial y \partial x}$ always. We will use the same example as before: $\frac{\partial^2 f}{\partial x \partial y} = 0$, and $\frac{\partial^2 f}{\partial y \partial x} = 0$.

Example 1.3

$f(x, y, t) = 2x^2y + \sin(\pi t)$, find $\frac{\partial f}{\partial x}, \frac{\partial f}{\partial y}, \frac{\partial f}{\partial t}, \frac{\partial^2 f}{\partial x^2}, \frac{\partial^2 f}{\partial y^2}, \frac{\partial^2 f}{\partial t^2}, \frac{\partial^2 f}{\partial x \partial y}$, and $\frac{\partial^2 f}{\partial y \partial x}$.

SOLUTION

$\frac{\partial f}{\partial x} = 4xy$, $\frac{\partial f}{\partial y} = 2x^2$, and $\frac{\partial f}{\partial t} = \pi \cos(\pi t)$

$\frac{\partial^2 f}{\partial x^2} = 4y$, $\frac{\partial^2 f}{\partial y^2} = 0$, and $\frac{\partial^2 f}{\partial t^2} = -\pi^2 \sin(\pi t)$; $\frac{\partial^2 f}{\partial x \partial y} = 4x$, and $\frac{\partial^2 f}{\partial y \partial x} = 4x$.

Gradient ∇

Gradient operator ∇ is a vector. We will recap some vector-related principles first.

If the displacement vector (three-dimensional) of fluid is $\boldsymbol{s} = x\vec{i} + y\vec{j} + z\vec{k}$, where ***i***, ***j***, and ***k*** are the unit vector in *x*, *y*, and *z* directions, respectively, then the values of *x*, *y*, and *z* are the components of the displacement in these three directions.

If *x*, *y*, and *z* change with time *t*, i.e. they are the functions of time, then the rate of change of the displacement with time is the velocity vector, **v**, of the fluid:

$$\mathbf{v} = \frac{\partial x}{\partial t}\vec{i} + \frac{\partial y}{\partial t}\vec{j} + \frac{\partial z}{\partial t}\vec{k} = v_x\vec{i} + v_y\vec{j} + v_z\vec{k} \tag{1.4}$$

where v_x, v_y, and v_z are the components of the velocity, **v**, in *x*, *y*, and *z* directions, respectively.

A dot product is a special operation for two vectors. The result of a dot product is not a vector but a scalar. For example, $\mathbf{a} \cdot \mathbf{v}$, where $\mathbf{a} = a_x\vec{i} + a_y\vec{j} + a_z\vec{k}$:

$$\mathbf{a} \cdot \mathbf{v} = a_x\frac{\partial x}{\partial t} + a_y\frac{\partial y}{\partial t} + a_z\frac{\partial z}{\partial t} = a_xv_x + a_yv_y + a_zv_z. \tag{1.5}$$

Most of the properties in fluid flow fields discussed in aerodynamics are the function of coordinates *x*, *y*, and *z* in space and time *t*. The change in any direction in space of any property can be examined by its partial derivative. The property can be, for example, temperature of fluid T, or density of fluid ρ, or the speed in x direction, v_x. The changes of the property in all directions can be expressed together by the ***gradient*** ∇ of the property.

The gradient is a vector operator:

$$\nabla = \frac{\partial}{\partial x}\vec{i} + \frac{\partial}{\partial y}\vec{j} + \frac{\partial}{\partial z}\vec{k} \tag{1.6}$$

For example, the gradient (change in 2-D space) of temperature T is:

$$\nabla T = \frac{\partial T}{\partial x}\vec{i} + \frac{\partial T}{\partial y}\vec{j}. \tag{1.7}$$

We can assume the temperature is the fluid temperature over a horizontal hot plate, the x direction is in parallel with the plate, and the y direction is perpendicular to the plate with $y=0$ at the plate surface. If the temperature on the plate is constant, $\frac{\partial T}{\partial x} = 0$, and the temperature above the plate ($y > 0$) is cooler than that on the plate, then $\frac{\partial T}{\partial y} < 0$.

Looking at another example, airflow over an aerofoil: assume x – chord-wise direction; y – vertical direction; and z – span-wise direction, and the gradient of the airspeed in x direction, v_x (in 3-D space), is:

$$\nabla v_x = \frac{\partial v_x}{\partial x}\vec{i} + \frac{\partial v_x}{\partial y}\vec{j} + \frac{\partial v_x}{\partial z}\vec{k} \tag{1.8}$$

At the front part above an aerofoil, airspeed increases along its chord: $\frac{\partial v_x}{\partial x} > 0$; air speed on the surface of the aerofoil is 0, and it gradually increases to the same level as that in free stream, so $\frac{\partial v_x}{\partial y} > 0$ above the aerofoil; in the span-wise direction, airspeed can decrease from wing root to wing tip: $\frac{\partial v_x}{\partial z} < 0$.

The dot product of velocity $\mathbf{v}$ and gradient ∇ is a common appearance in fluid-dynamic equations when the equations describe the total change of properties of fluid with respect to time t. The result of the dot product $\mathbf{v} \cdot \nabla$ is:

$$\mathbf{v} \cdot \nabla = v_x \frac{\partial}{\partial x} + v_y \frac{\partial}{\partial y} + v_z \frac{\partial}{\partial z} \tag{1.9}$$

For example: $v_x \frac{\partial T}{\partial x} + v_y \frac{\partial T}{\partial y} + v_z \frac{\partial T}{\partial z}$ can be rewritten as $\mathbf{v} \cdot \nabla T$

and $\mathbf{v} \cdot \nabla v_x = v_x \dfrac{\partial v_x}{\partial x} + v_y \dfrac{\partial v_x}{\partial y} + v_z \dfrac{\partial v_x}{\partial z}$.

Total derivative

A partial derivative can describe the changes of a multi-variable function with respect to a specific variable, as discussed above. However, the total change, or complete change of the function with respect to a specific variable, can be analysed by a total/complete differentiation of the function. This total/complete differentiation will consider the changes of the function related to all variables, and consider all of the variables are sub-functions of the specific variable.

x, y and t are the variables of function $f(x, y, t)$. To carry out the total differentiation of this function with respect to t, we consider that x and y are functions of t, applying the *Chain Rule*:

$$\frac{df}{dt} = \frac{\partial f}{\partial t} + \frac{\partial f}{\partial x}\frac{dx}{dt} + \frac{\partial f}{\partial y}\frac{dy}{dt} \tag{1.10}$$

Using the same method, the total derivative with respect to x (treat y and t as functions of x) is:

$$\frac{df}{dx} = \frac{\partial f}{\partial t}\frac{dt}{dx} + \frac{\partial f}{\partial x} + \frac{\partial f}{\partial y}\frac{dy}{dx}, \tag{1.11}$$

and the total derivative with respect to y (treat x and t as functions of y) is:

$$\frac{df}{dy} = \frac{\partial f}{\partial t}\frac{dt}{dy} + \frac{\partial f}{\partial x}\frac{dx}{dy} + \frac{\partial f}{\partial y}. \tag{1.12}$$

From equations (1.10), (1.11), and (1.12), we can obtain the total change of the function. It can be expressed as:

$$df = \frac{\partial f}{\partial t}dt + \frac{\partial f}{\partial x}dx + \frac{\partial f}{\partial y}dy \tag{1.13}$$

For example, for the function used in ***Example 1.3***: $f(x, y, t) = 2x^2y + \sin(\pi t)$, the total change of this function is:

$$df = \pi\cos(\pi t)dt + 4xydx + 2x^2dy.$$

In fluid mechanics and aerodynamics, a property, for example, temperature, or velocity/speed of fluid particles, is a function of the location of fluid particles, i.e. in x, y, and z coordinate space, and x, y, and z are functions of time t. Therefore, the total derivative of a property f with respect to time t is:

$$\frac{df}{dt} = \frac{\partial f}{\partial t} + \frac{\partial f}{\partial x}\frac{dx}{dt} + \frac{\partial f}{\partial y}\frac{dy}{dt} + \frac{\partial f}{\partial z}\frac{dz}{dt} \tag{1.14}$$

and $\frac{dx}{dt}, \frac{dy}{dt}$, and $\frac{dz}{dt}$ are the change rates of displacement in x, y, and z directions, i.e. the velocity components in x, y, and z directions, respectively. Therefore, the total derivative with respect to t is:

$$\frac{df}{dt} = \frac{\partial f}{\partial t} + v_x\frac{\partial f}{\partial x} + v_y\frac{\partial f}{\partial y} + v_z\frac{\partial f}{\partial z} = \frac{\partial f}{\partial t} + \mathrm{v}\cdot\nabla f \tag{1.15}$$

This means that the total change of a property is the sum of its temporal (t) change $\frac{\partial f}{\partial t}$ and its spatial change, $v_x\frac{\partial f}{\partial x} + v_y\frac{\partial f}{\partial y} + v_z\frac{\partial f}{\partial z} = \mathrm{v}\cdot\nabla f$.

For example, the total change rate of mass m of fluid with respect to time is:

$$\frac{dm}{dt} = \frac{\partial m}{\partial t} + v_x\frac{\partial m}{\partial x} + v_y\frac{\partial m}{\partial y} + v_z\frac{\partial m}{\partial z}. \tag{1.16}$$

The total change rate of airspeed in the x direction, v_x, is:

$$\frac{Dv_x}{Dt} = \frac{dv_x}{dt} = \frac{\partial v_x}{\partial t} + v_x\frac{\partial v_x}{\partial x} + v_y\frac{\partial v_x}{\partial y} + v_z\frac{\partial v_x}{\partial z} = \frac{\partial v_x}{\partial t} + \mathbf{v}\cdot\nabla v_x. \tag{1.17}$$

Integration

Integration is the other part of calculus. Integration is the inverse function process of differentiation, called anti-differentiation. The symbol of integrating a function $f(x)$ is:

$$\int f(x)dx,$$

which is also called integral.

If the derivative of $F(x)$ is $f(x)$, the process to obtain $f(x)$ from $F(x)$ or vice versa is:

$$F(x) \underset{\text{Integration}}{\overset{\text{Differentiation}}{\rightleftarrows}} f(x)$$

There are two types of integration: indefinite integral and definite integral.

Indefinite integral

Differentiation rules shows that
if the derivative of a function $F(x)$ is $f(x)$: $F'(x)=f(x)$, $F_1(x)= F(x)+1$ and $F_2(x)= F(x)-2$,
then $F_1'(x)=f(x)$, and $F_2'(x)=f(x)$ as well,
i.e. the derivative of $F(x)+C$ is $f(x)$, where C is a constant.
It is fair to say that $f(x)$ can be the derivative of a group of functions.
Therefore,

$$\int f(x)dx = F(x)+C \tag{1.18}$$

is called the ***indefinite integral***. It represents a group of functions, parallel to each other.

Example 1.4

Find indefinite integrals:

1) $f(x)=0.3$.

$$y=\int 0.3dx = 0.3x+C$$

The result is a group of parallel lines, as shown in Figure 1.5.

2) $f(x)=x$.

$$y=\int xdx=\frac{1}{2}x^2+C,\ F(x)=\frac{1}{2}x^2$$

The result is a group of parabolic lines, as shown in Figure 1.6.

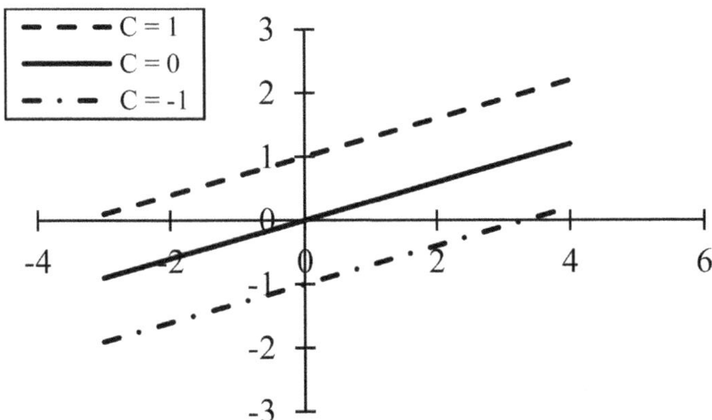

FIGURE 1.5
A group of linear functions as the result of an indefinite integral: $\int 0.3dx$.

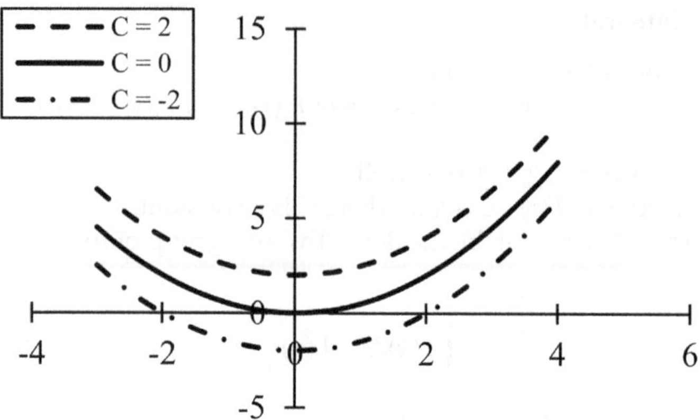

FIGURE 1.6
A group of quadratic functions as the result of an indefinite integral: $\int xdx$.

Definite integral

Integration with two limits, $\int_a^b f(x)dx$, is called the ***definite integral***, where a is the lower limit and b is the upper limit of the integration. The definite integral returns a value which can have a physical unit depending on the function.

The process for finding a definite integral is similar to that for an indefinite integral: integrate $f(x)$ to find $F(x)$, and then substitute $x=a$ and $x=b$ into $F(x)$ to find the difference:

$$\int_a^b f(x)dx = F(b) - F(a). \tag{1.19}$$

If $y=f(x)$ is a curve in x–y coordinates, $\int_a^b f(x)dx$ is the area under the curve between $x=a$ and $x=b$.

Example 1.5

$f(x)=x$, find the definite integral: $\int_0^2 xdx = \frac{1}{2}x^2$

$$\int_0^2 xdx = \frac{1}{2}x^2\Big|_0^2 = 0.5\times 2^2 - 0.5\times 0 = 2$$

The result can be described by the shaded area under the straight line $f(x)=x$ between $x=0$ and $x=2$, as shown in Figure 1.7.

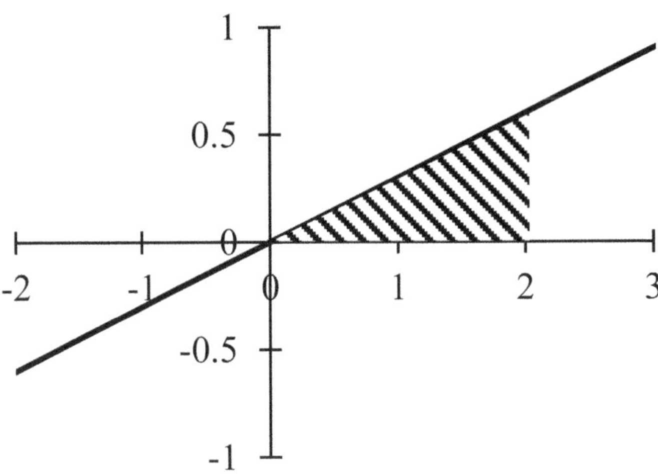

FIGURE 1.7

Graphic illustration of a definite integral: $\int_0^2 x dx$.

Exercises

1. Find the extreme value of $f(x) = x^2 + 1$.
2. For function $f(x, y, z) = xyz$, find $\frac{\partial f}{\partial x}, \frac{\partial f}{\partial y}, \frac{\partial f}{\partial z}$ and df.
3. Integrate $\int dp + \int v dv$.

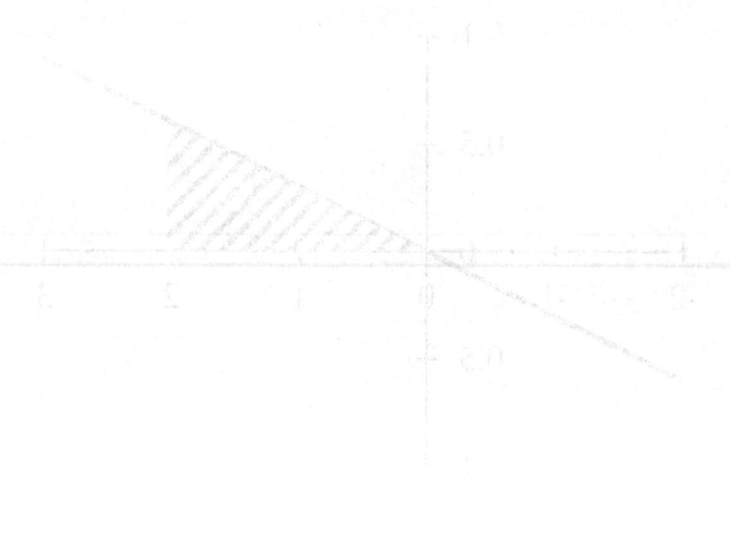

2

Fundamental Principles of Aerodynamics (Subsonic)

Aerodynamics principles explain the properties and functions of the airflow around an aircraft. When air flows around an aircraft, its property changes. This change would follow different "rules" if the aircraft flies at different airspeed. The focus of this chapter is at the aerodynamic principles when airspeed is relatively low – in subsonic region. The fundamental knowledge you gained in physics is the starting point to build aerodynamic principles. There are a few of fundamental principles of physics used in aerodynamics discussed in this book:

- Conservation of mass
- Newton's laws of motion
- Conservation of energy
- Ideal gas law

From these principles, we will derive the following basic and often-used equations: Continuity equation, Bernoulli's equation, and Energy equation (of fluid flow). Those equations describe the behavior of fluid flow along its path and the changes of fluid property, i.e. pressure, density, and temperature in the flow. We will adopt the *Eulerian viewpoint*, which is to observe a fluid particle along a streamline at a particular moment of time, to study the pattern/behavior of fluid flow.

A *Streamline* is the line, which shows the direction of a fluid flow at any particular moment. Streamlines do not intercept. It indicates that the speed of a fluid flow is relatively high when the streamlines are drawn closer to each other; the speed is relatively low if the streamlines are apart from each other.

Continuity Equation

The conservation of mass in physics states that the mass in a defined system cannot be created, or destroyed, if there is not a source (any supply of mass) or a sink (any consumption of mass) within this system. In the other words, the total change of mass is zero in this system. Figure 2.1 shows a defined system: a pipe flow, in which (between streamlines) a fluid flows in from left side, section 1, and flows out from the right side, section 2 of the system. In

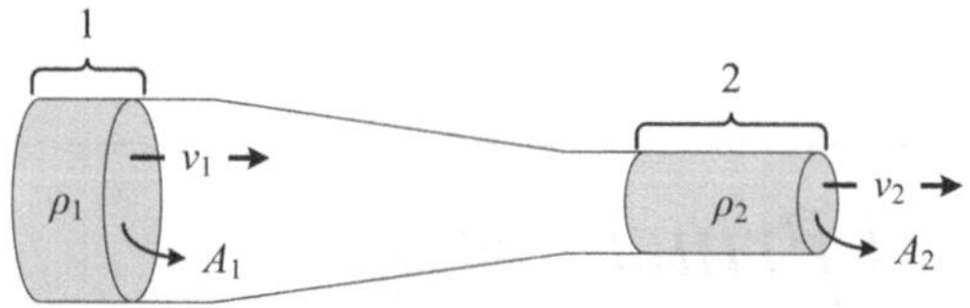

FIGURE 2.1
Fluid flow through an uneven diameter pipe.

this piece of pipe, there is no new mass being created and there is no mass being destroyed, so the total change of mass should be zero, i.e. the total mass within the system is constant. Therefore, the mass flowing-in is equal to the mass flowing-out. That is, the mass flow rate at section 1 is the same as the mass flow rate at section 2.

Mass flow rate is $\dot{m} = \rho v A\,[\text{kgs}^{-1}]$, where A is the section area in m^2; v: average speed of the fluid at this section in ms^{-1}; and ρ is the density of the fluid in kgm^{-3}. So the *Continuity equation,* which describes the conservation of mass, is:

$$\rho_1 v_1 A_1 = \rho_2 v_2 A_2 \tag{2.1}$$

i.e. *the mass flowrate of a fluid between streamlines is constant if there is no mass source or mass sink between the streamlines.*

If the density of the fluid does not change in the system, $\rho_1 = \rho_2$, i.e. the fluid is *incompressible,* the Continuity equation becomes

$$v_1 A_1 = v_2 A_2 \tag{2.2}$$

Equation (2.2) state that the volumetric flowrate of an incompressible fluid is constant in this system as well (volumetric flowrate $= vA$ $[\text{m}^3\text{s}^{-1}]$).

Example 2.1

Oil flows through a piece of pipe, whose diameter changes. At the section of inlet of the pipe, the diameter $\phi_1 = 0.02$ m, oil flow speed v_1 is 2 m/s. At the outlet of the pipe, the diameter becomes $\phi_2 = 0.005$ m. Oil can be treated as an incompressible fluid, with a density of 850 kgm^{-3}.

1. What is the oil speed at the outlet v_2?
2. What is the mass flowrate of oil in the pipe?

SOLUTION

1. Oil is incompressible. Use Equation (2.2):

$$v_2 = \frac{v_1 A_1}{A_2} = \frac{2 \times \pi \times \left(\frac{0.02}{2}\right)^2}{\pi \times \left(\frac{0.005}{2}\right)^2} = 32\ \text{ms}^{-1}.$$

2. Mass flowrate $\dot{m} = \rho_1 v_1 A_1 = 850 \times 2 \times \pi \left(\frac{0.02}{2}\right)^2 = 0.534\ \text{kgs}^{-1}.$

Bernoulli's Equation

Considering the forces exerted on a fluid particle of m (mass) in the fluid flow along a streamline s as shown in Figure 2.2, A is the average area of cross-section of the fluid particle; δs is the length of the particle along the streamline; δh is elevation difference between the ends of the particle, and $\sin\alpha = \dfrac{\delta h}{\delta s}$ in Figure 2.2; p is local pressure of the fluid, and δp is the pressure change along the streamline; and v is the velocity of the particle. The mass of the fluid particle is:

$$m = \rho A \delta s. \tag{2.3}$$

The forces exerted on this particle are gravitational force and the force from the pressure on the particle, if we ignore the friction (none-viscous flow). We apply Newton's Second Law of Motion, $\Sigma F = ma$, to this particle along the streamline. ΣF is the total force exerted on the particle, and a is the acceleration.

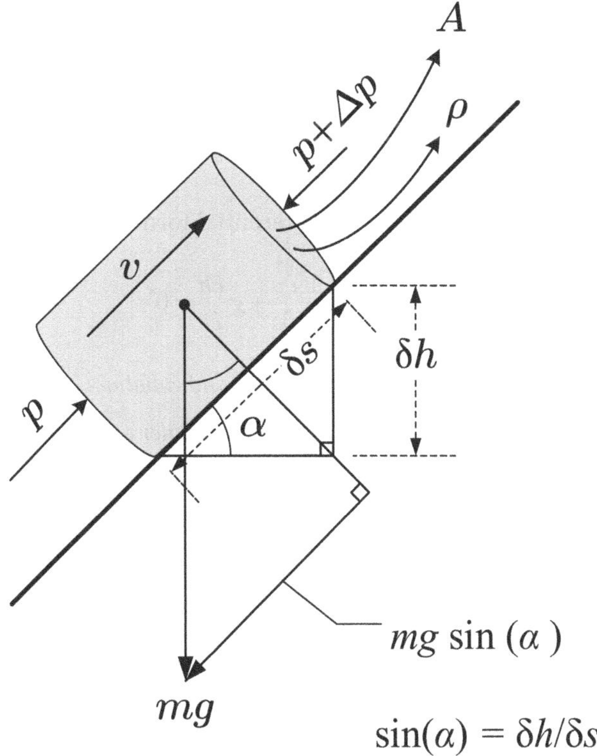

FIGURE 2.2
A fluid particle in streamline.

The forces due to the pressure on the particle are pA, and $-(p+\delta p)A$, and the component of gravitational force along the streamline is $-mg\dfrac{\delta h}{\delta s}$, where g is the gravitational acceleration ($g=9.8$ ms^{-2} used in this book). So the total force on the particle is:

$$\Sigma F = A(p-(p+\delta p)) - mg\frac{\delta h}{\delta s} \tag{2.4}$$

The expression for acceleration a of the fluid particle in a fluid field is a total derivative of velocity v with respect to time t as described in Chapter 1 (1.15): $a=\dfrac{Dv}{Dt}=\dfrac{dv}{dt}=\dfrac{\partial v}{\partial t}+v\nabla v.$

If we consider the flow along a streamline s only (one-dimensional), the total derivative becomes:

$$a=\frac{dv}{dt}=\frac{\partial v}{\partial t}+v\frac{\partial v}{\partial s}. \tag{2.5}$$

Substitute Equation (2.3), (2.4), and (2.5) into the expression of Newton's Second Law of Motion of unit mass of fluid particle along the streamline, we can obtain:

$$\frac{\partial v}{\partial t}+v\frac{\partial v}{\partial s}=\frac{1}{\rho A\delta s}\left((p-p-\delta p)A-g\rho A\delta s\frac{\delta h}{\delta s}\right) \tag{2.6}$$

Then simplify and rewrite (2.6) in a differential form:

$$\frac{\partial v}{\partial t}+v\frac{\partial v}{\partial s}+\frac{1}{\rho}\frac{\partial p}{\partial s}+g\frac{\partial h}{\partial s}=0 \tag{2.7}$$

Equation (2.7) is known as the *Euler (momentum) equation.*

If the fluid flow is in a *steady state,* $\dfrac{\partial}{\partial t}=0$, the *Euler* equation (2.7) becomes

$$g\frac{dh}{ds}+\frac{1}{\rho}\frac{dp}{ds}+v\frac{dv}{ds}=0 \tag{2.8}$$

which is a function of streamline only, and can be integrated along a streamline:

$$\int\left(g\frac{dh}{ds}+\frac{1}{\rho}\frac{dp}{ds}+v\frac{dv}{ds}\right)ds=C \tag{2.9}$$

If the fluid is *incompressible,* the integral (2.9) becomes:

$$gh+\frac{p}{\rho}+\frac{v^2}{2}=C \tag{2.10a}$$

or:

$$gh\rho+p+\frac{\rho}{2}v^2=C \tag{2.10b}$$

where h is elevation, height, in m; p is static pressure of the fluid in Pa; $g \approx 9.8$ ms^{-2}, gravitational acceleration; and C is a constant. Equation (2.10) is called *Bernoulli's equation*. The conditions required to obtain this equation can be summarized as follows: incompressible fluid particle flows along a streamline; the flow is in steady state; and consider no friction.

Each term of the equations represents a type of energy per unit mass (2.10a), or per unit volume (2.10b) of the fluid along the streamline. The statement of the Bernoulli's equation is that for a flow of an incompressible ideal fluid along its streamline the sum of the potential energy ($gh\rho$), kinetic energy ($\rho(v^2/2)$) and the capability (p) to produce work due to pressure at any point on the streamline remains constant – *Bernoulli's Theorem*.

Take the fluid flow shown in Figure 2.1 as an example, there are two labeled sections: upstream section 1 and downstream section 2. p_1, and p_2 are the pressure at these two sections, v_1, and v_2 are the speed of the fluid, and h_1, and h_2 are the height at these two sections respectively. According to Bernoulli's Theorem, the sum of the energies mentioned above remains constant; therefore, we can obtain:

$$gh_1\rho+p_1+\frac{\rho}{2}v_1^2=gh_2\rho+p_2+\frac{\rho}{2}v_2^2 \tag{2.11}$$

From Equation (2.11) we learn that the speed of the fluid flow will decrease if the fluid flows up if the fluid pressure is the same at both sections; or when flow speed increases, the pressure will decrease if these two sections are at the same level.

Each term of Bernoulli's Equation (2.11) has the unit of pressure, [Pa]. We can use P as the constant in Equation (2.10):

$$gh\rho+p+\frac{\rho}{2}v^2=P \tag{2.12}$$

in which,

$gh\rho$ – *"Potential" pressure;* p – *Static pressure;*

$\frac{\rho}{2}v^2$ – *Dynamic pressure;* P – *Total pressure.* (Sometimes p_t is used)

Divide each term of Equation (2.12) by $g\rho$, assuming the fluid is incompressible, and it becomes:

$$h+\frac{p}{g\rho}+\frac{v^2}{2g}=H \tag{2.13}$$

The terms of Equation (2.13) have the unit of length/height, [m].

In Equation (2.13),

h – (*"Natural" height) Head;* $\frac{p}{g\rho}$ – *Static head;*

$\frac{v^2}{2g}$ – *Dynamic head;* H – *Total head.*

The expression of "heads" are commonly used in engineering fluid mechanics to describe the capacity of pumps, fluid flow in pipes and ducts.

In our application of analyzing airflow around an aircraft, air is assumed as an incompressible fluid if the aircraft speed is less than 250 kt (=128.5 ms^{-1}. 1 kt = 0.514 ms^{-1}), and it is commonly assumed that h in the Bernoulli's equation (2.12) is constant. So the Bernoulli's equation to the airflow around a low subsonic aircraft becomes:

$$p+\frac{\rho}{2}v^2=P \tag{2.14}$$

This means that the sum of static pressure and dynamic pressure is the total pressure, which is constant.

Example 2.2

A level flow: air pressure $p_1=1.035\times10^5$ Pa, v_1 = 50 m/s, ρ = 1.125 kg/m^3 (Assume: air is incompressible here.), and v_2 = 70 m/s.

Calculate

1. p_2
2. total pressure P
3. dynamic pressure p_d at v_1 = 50 m/s
4. dynamic head h_d at v_2 = 70 m/s

SOLUTION

1. Use Bernoulli's equation (2.11):

$$p_2+\frac{\rho}{2}v_2^2+h=p_1+\frac{\rho}{2}v_1^2+h$$

$$p_2=1.035\times10^5+\frac{1.125}{2}\times50^2-\frac{1.125}{2}\times70^2$$

$$p_2=1.0215\times10^5\ \text{(Pa)}$$

2. Total pressure P by Equation (2.14):

$$P = p + \frac{\rho}{2}v^2$$

$$P = 1.035 \times 10^5 + \frac{1.125}{2} \times 50^2 = 1.049 \times 10^5 \text{ (Pa)}$$

3. $p_d = \frac{\rho}{2}v_1^2 = \left(\frac{1.125}{2}\right) \times 50^2 = 1406.25 \text{ (Pa)}$

4. $h_d = \frac{1}{2g}v_2^2 = \left(\frac{1}{2 \times 9.8}\right) \times 70^2 = 250 \text{ (m)}.$

Stagnation Pressure

According to Bernoulli's equation, static pressure of airflow increases with the decreasing speed along a streamline. When the air speed decreases to "0", for example, at the leading edge of aerofoil, its static pressure reaches the highest value. The location, where the air speed decreases down to "0", is called *stagnation point,* and the static pressure, density and temperature of air at this point are called stagnation pressure, stagnation density, and stagnation temperature respectively. For a level fluid flow, the static stagnation pressure, p_{stg}, is equal to the total pressure P (p_t).

Example 2.3

An airflow travels at $v_1 = 25$ ms^{-1}, $p_1 = 1.03 \times 10^5$ Pa, then its speed increases to $v_2 = 32$ ms^{-1}. Its density $\rho = 1.225$kgm^{-3}. Find p_2, and the stagnation pressure p_s.

SOLUTION

32 ms^{-1} = 16.45 kt, so air is incompressible. Use Bernoulli's equation:

$$p_2 + \frac{\rho}{2}v_2^2 = p_1 + \frac{\rho}{2}v_1^2$$

$$p_2 = 1.03 \times 10^5 + \frac{1.225}{2} \times 25^2 - \frac{1.225}{2} \times 32^2$$

$$p_2 = 1.028 \times 10^5 \text{ (Pa)}$$

The stagnation pressure p_{stg}:

$$p_{stg} + 0 = p + \frac{\rho}{2}v^2 = P$$

$$p_{stg} = 1.03 \times 10^5 + \frac{1.225}{2} \times 25^2 = 1.034 \times 10^5 \text{ (Pa)}$$

Applications of Bernoulli's Equation

For a subsonic flight whose airspeed normally does not exceed 250 knots, air can be treated as an incompressible fluid. Bernoulli's equation is one of the fundamental concepts for analyzing the dynamic forces and air properties around the aircraft. For example, to measure airflow rate to the aircraft engine, the airspeed of an aircraft, analyze the lift produced by an aerofoil and so on.

Venturi Tube (Flowrate Meter)

The Venturi effect is the fact that reduction of static pressure fluid is caused by decreasing the cross-section area of the flow path. Examples of the Venturi effect are shown in Figure 2.3: (a) the airflow between two moving aircraft; (b) Venturi tube; and (c) carburetor of a petrol engine.

A structure consisting of a horizontal tube with two different cross-section areas, as shown in Figure 2.3(b) and Figure 2.4, is used to measure flow rates of incompressible fluid. This is called Venturi tube or Venturi meter.

Figure 2.4 shows the principle of measuring flow rate of incompressible fluid. An incompressible fluid flows through the narrow section 2, and its

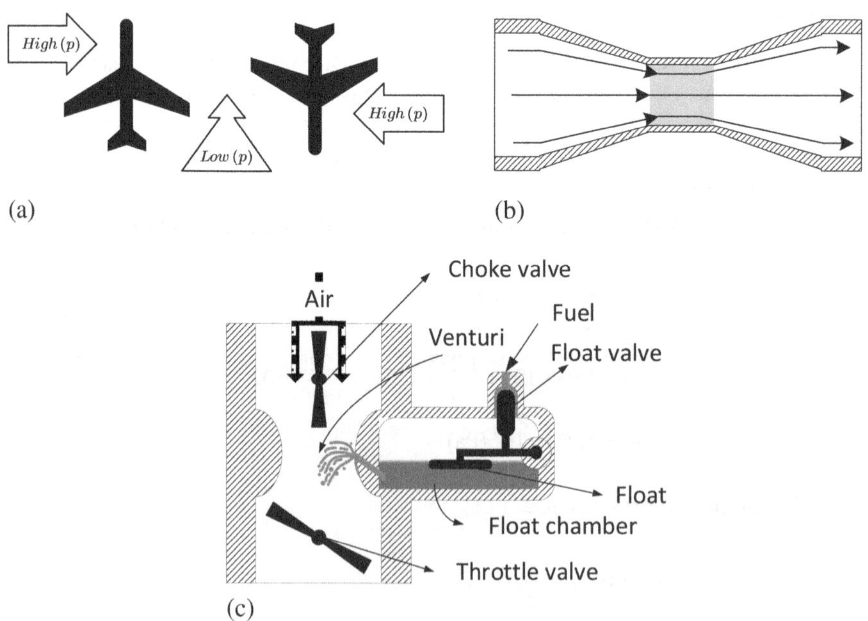

FIGURE 2.3
Examples of Venturi effect: (a) airflow between two airplanes, (b) Venturi tube, (c) carburetor.

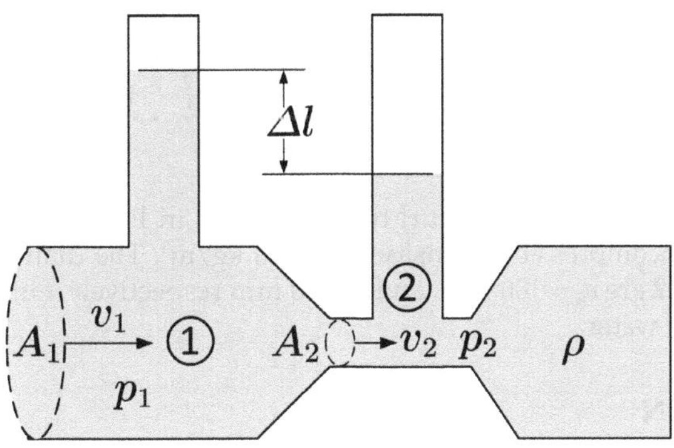

FIGURE 2.4
Venturi tube.

speed increases $v_2 > v_1$, according to the Continuity equation (2.2). Examine these two sections by using Bernoulli's equation:

$$gh_1\rho + p_1 + \frac{\rho}{2}v_1^2 = gh_2\rho + p_2 + \frac{\rho}{2}v_2^2 \tag{2.15}$$

$h_1 = h_2$, because it is level in horizontal. So the difference between v_1 and v_2 causes the difference in static pressure between section 1 and section 2. By reorganizing (2.15), we obtain:

$$\Delta p = (p_1 - p_2) = \frac{\rho}{2}\left(v_2^2 - v_1^2\right) \tag{2.16}$$

Δp is measured by Δl, the difference of the heights of the fluid (static heads) in the vertical pipes at the section 1 and section 2 (Figure 2.4):

$$\Delta p = \rho g \Delta l. \tag{2.17}$$

where Δl is the difference of static heads in m.

Using the Continuity equation (2.2), $v_2 = \frac{A_1}{A_2}v_1$, where A_1 and A_2 are the area at section 1 and section 2, respectively, substitute the v_2 and Equation (2.17) into Equation (2.16):

$$\rho g \Delta l = \frac{\rho}{2}\left(\frac{A_1^2}{A_2^2}v_1^2 - v_1^2\right) = \frac{\rho v_1^2}{2}\left(\frac{A_1^2}{A_2^2} - 1\right)$$

then

$$v_1 = \sqrt{2g\Delta l \frac{A_2^2}{\left(A_1^2 - A_2^2\right)}}. \tag{2.18}$$

The mass flow rate is $\dot{m} = A_1\rho_1 v_1 = A_1\rho_1\sqrt{2g\Delta l\dfrac{A_2^2}{\left(A_1^2 - A_2^2\right)}}$, [kgs^{-1}].

Example 2.4

Water flows through a Venturi tube as shown in Figure 2.4: $\Delta l = 0.5$ m. Water is incompressible, density $\rho_w = 1000$ kg/m^3. The diameter of section 1 and 2 are $d_1 = 100$ mm, and $d_2 = 50$ mm respectively. Find the mass flowrate of water.

SOLUTION

To calculate the mass flowrate, the Continuity equation is used: $\dot{m} = \rho A_1 v_1 = \rho A_2 v_2$, and Bernoulli's equation is used to produce Equation (2.18). $A_1 = \pi\left(\dfrac{d_1}{2}\right)^2$ and $A_2 = \pi\left(\dfrac{d_2}{2}\right)^2$. So v_1 can be calculated:

$$v_1 = \sqrt{2g\Delta l\frac{A_2^2}{\left(A_1^2 - A_2^2\right)}} = A_2\sqrt{\frac{2\times 9.8\times 0.5}{3.14^2\times(0.05^4 - 0.025^4)}}$$

$$\dot{m} = A_1\rho_1 A_2\sqrt{2g\Delta l\frac{1}{\left(A_1^2 - A_2^2\right)}}$$

$$= 1000\times 3.14^2\times 0.05^2\times 0.025^2\sqrt{\frac{2\times 9.8\times 0.5}{3.14^2\times\left(0.05^4 - 0.025^4\right)}}$$

$$= 6.35\ (\text{kgs}^{-1})$$

The mass flowrate is 6.35 kgs^{-1}.

Pitot Tube (Airspeed)

The airspeed of an aircraft is commonly obtained through a Pitot tube. Figure 2.5 (a) is the schematic diagram of a Pitot tube, and Figure 2.5 (b) shows Pitot tubes used on a Jet-Stream 32. There are two channels within the Pitot tube (Figure 2.5 (a)): one is in the center, open to the airflow, where air becomes stagnant. Therefore, it measures the stagnation (total) pressure p_t of the air flow. The opening of the other channel is parallel to the airflow (shaded), which measures the static pressure p_s. The output of the Pitot tube is the pressure difference, i.e. the dynamic pressure, measured by those two channels:

$$p_d = p_t - p_s = \frac{1}{2}\rho v^2 \tag{2.19}$$

Then airspeed v can be calculated.

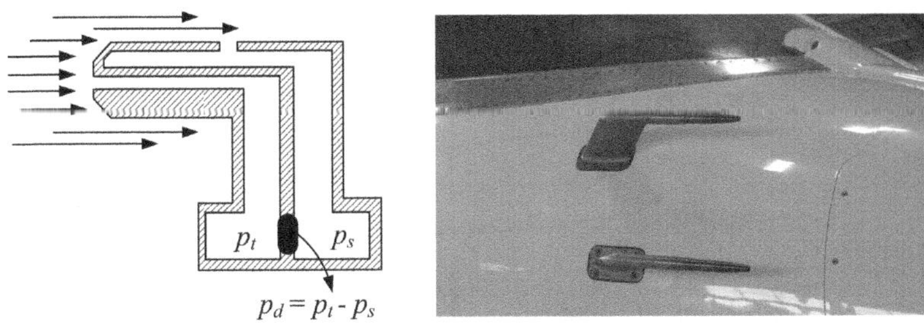

FIGURE 2.5
Pitot tube: (a) schematic diagram of Pitot tube, (b) Pitot tubes on a Jet-Stream 32.

Example 2.5

If a Pitot tube measures the difference between the static pressure: $p_s = 1.031 \times 10^5$ Pa, and the total pressure $p_t = 1.039 \times 10^5$ Pa. The air can be treated as incompressible, and its density: $\rho = 1.225$ kg/m^3.

(a) What is the dynamic pressure measured by the Pitot tube?
(b) What is the airspeed?
(c) What is the dynamic pressure if airspeed is 50 m/s?

SOLUTION

(a) The dynamic pressure, using Equation (2.19): $p_d = p_t - p_s = (1.039 - 1.031) \times 10^5 = 800$ Pa;

(b) The air speed can be calculated by Equation (2.19):

$$p_d = \frac{1}{2}\rho v^2, \quad v = \sqrt{\frac{2}{\rho} p_d} = \sqrt{\frac{2 \times 800}{1.225}} = 36.14 \text{ ms}^{-1};$$

(c) If $v = 50$ m/s,

$$p_d = \frac{1}{2}\rho v^2 = \frac{1}{2} \times 1.225 \times 50^2 = 1531 \text{ Pa}$$

The dynamic pressure is 1531 Pa.

Lift Produced by a Subsonic Aerofoil

When air flows around an aerofoil as shown in Figure 2.6, (with a small angle of attack, α), the relative airspeed, v, over the aerofoil increases, and the static pressure over the aerofoil will decrease according to Bernoulli's equation, which is indicated by the "–" over the aerofoil in Figure 2.6 (RA: relative airspeed). At the same time, the area of the flow path under the lower surface of the aerofoil increases, air speed decreases, and, in turn, the static pressure increases, which is indicated by "+".

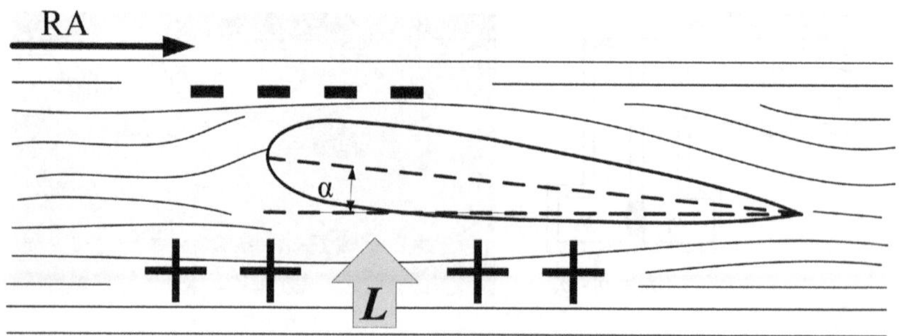

FIGURE 2.6
Aerofoil in a level airflow.

Due to the static pressure difference between the upper and lower surfaces, integrating the static pressure over the surface of the aerofoil results in an upward total force in vertical direction exerted on the aerofoil, which is the lift, labeled L in Figure 2.6.

Ideal Gas Law

An equation (mathematical expression) that describes the relationships of properties, p (pressure) in Pa, V (volume) in m^3, and T (temperature) in K, of an ideal gas, is called the *Ideal Gas Law (IGL)*:

$$pV = nRT \tag{2.20}$$

where R is the *universal gas constant*, 8.314 J(molK)$^{-1}$, and n is the number of moles of the gas. $n = \frac{m}{M}$, where m is the mass in grams of the gas in volume V, while M is the molar mass in grams of the gas. The expression for n in (2.20) can be rearranged as:

$$pV = nRT = \frac{m}{M}RT = m\frac{R}{M}T$$

Then:

$$p\frac{V}{m} = \frac{R}{M}T \Rightarrow \frac{p}{\rho} = R_M T \tag{2.21}$$

Equation (2.21) is another form of the Ideal Gas Law, where ρ is the density of the gas, in kgm^{-3}, pressure p is in Pa; temperature T is absolute temperature in K (0°C = 273.15 K, and 273 K is used in this book); and R_M is the *gas constant* in J(kgK)$^{-1}$, whose value depends on the gas. For example, air can be treated as an ideal gas, and its molar mass is approximate 28.97 g, so the gas constant of air is 287 J(kgK)$^{-1}$.

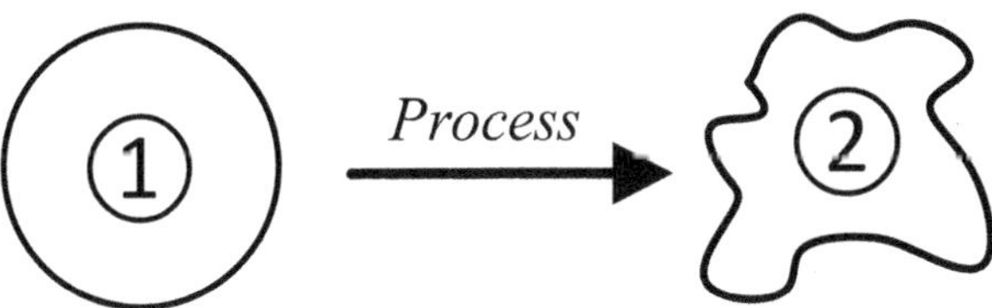

FIGURE 2.7
State change of a gas system.

The Ideal Gas Law deals with the property of a gas, pressure, density, and temperature, which define the state of the gas (Figure 2.7). The Ideal Gas Law is also *State Equation of Gas*. If an ideal gas system is in a state of p_1, ρ_1, and T_1, and then the state of the gas changes, for example, by heating or cooling or compression, the pressure, density and temperature reach a set of new values, p_2, ρ_2, and T_2., as shown in Figure 2.7. The process should be reversible, which means that the change within the system is uniform and can be reversed completely. Then, both sets of p_1, ρ_1, and T_1 and p_2, ρ_2, and T_2 should satisfy the Ideal Gas Law (2.21):

$$\frac{p_1}{\rho_1 T_1} = \frac{p_2}{\rho_2 T_2} = R_M \tag{2.22}$$

Example 2.6

Find the air density ρ, if air temperature T is 15°C, and air pressure p is 1.031×10^5 Pa. If the air temperature decreases to 0°C, and the density is down to 1.2 kgm^{-3}, what is its pressure now?

SOLUTION

$T = 15 + 273 = 288$ K. Use Ideal Gas Law – State equation (2.21), and $R_M =$ 287 J(kgK)$^{-1}$ for air:

$$\frac{p}{\rho} = R_M T \Rightarrow \rho = \frac{p}{R_M T} = \frac{1.013 \times 10^5}{287 \times 288} = 1.225 \text{ (kgm}^{-3}\text{)}$$

The density at this condition is 1.225 kgm^{-3} (sea level condition).

When it the temperature decreases to $T_2 = 0 + 273 = 273$ K, and $\rho_2 = 1.2$ kgm^{-3}, we can use IGL either (2.21) or (2.22) to calculate p_2. We use (2.22) without the gas constant R_M to find p:

$$\frac{p}{\rho T} = \frac{p_2}{\rho_2 T_2} \Rightarrow p_2 = \frac{\rho_2 T_2}{\rho T} p$$

$$= \frac{1.2 \times 273}{1.225 \times 288} \times 1.013 \times 10^5$$

$$= 9.406 \times 10^4 \text{ (Pa)}$$

First Law of Thermodynamics

A gas system can receive heat from outside to raise its own temperature; for example, the temperature of a bottle of gas sitting in the sun increases. This system can also receive work from external sources to increase its pressure, or the gas system expands to push a load to produce work, like an internal combustion engine's cylinder. All of these processes involve energy transfer. In any of the processes, *the heat energy received by a gas system leads to the internal energy increase of the gas in the system, and the work the system does to outside of the system if this gas system is a closed system* (there is a defined amount of mass, *m*, within the system) – this statement is called the *First Law of Thermodynamics* (Energy Conservation):

$$Q = \Delta E_{\text{int}} + W \tag{2.23}$$

where Q is the heat received by the system; ΔE_{int} is the change of internal energy of the gas; and W is work done by the system. "+" Q means that the system receives heat, and "−" Q, loses heat. "+" ΔE_{int} means that the internal energy of the system increases, and "−" ΔE_{int}, internal energy decreases. "+" W means that the system does work to others, and "−" W, receives work from outside the system.

The differential form of the First Law of Thermodynamics is

$$dQ = dE_{\text{int}} + dW \tag{2.24}$$

If the gas in this system is ideal gas, from the kinetic theory of gases (Serway and Jewett, 2008), the internal energy of ideal gas is a function of temperature only: $dE_{\text{int}} = d(mC_V T)$. m is the mass of the gas in the system. C_V is the specific heat [J(kgK)$^{-1}$] of the gas when its volume (density) is constant, so $dE_{\text{int}} = mC_V dT$.

In thermophysics, in the microscopic scale, the work done by gas of unit mass can be expressed as the product of pressure and its volume change: $dW = pdV$.

Therefore, the First Law of Thermodynamics (2.24) can be rewritten as:

$$dQ = mC_V dT + pdV \tag{2.25}$$

Example 2.7

Analyze the following processes by First Law of Thermophysics (2.23):

1. Heating up water in a pot (not boiling)
2. Compressing air in a glass syringe quickly
3. Gas bottle under the sun

4. Heating steam to push a piston with a constant load.
5. Compress gas in a syringe by the plunger while the whole system is in ice+water.

SOLUTION

1. Q: +; ΔE_{int}: +; $W = 0$.
2. $Q = 0$; ΔE_{int}: +; W: –.
3. Q: +; ΔE_{int}: +; $W = 0$.
4. Q: +; ΔE_{int}: +; W: +.
5. Q: –; $\Delta E_{int} = 0$; W: –.

Processes

Gases can be used as media in many thermal processes to produce power or to generate various forces, or to transfer energy: gases in an engine, in a boiler, or a refrigerator, or the air flowing around an aerofoil, propellers, blades of a compressor or turbine, and so on. These processes are complicated, and the state of gases in these processes change from one state to another. The functions of two different complex processes can be different, even if the two processes have the same initial state and the same final state of the gas. However, any application of thermal processes can be described as a combination of one or more basic simple processes. There are four basic thermal processes. Each of them progresses under a specific condition.

p-V Diagram

In order to analyze a thermal process clearly and conveniently with a visual impression, a process can be displayed in a diagram called a *p-V* diagram: the volume V of the system is in the horizontal axis, while the pressure p is the vertical axis, as shown in Figure 2.8. In this *p-V* diagram, the system starts at the initial state at point "1" (p_1, V_1), and then changes to state "2" (p_2, V_2). The path/curve between "1" to "2" is the thermal process the system follows from "1" to "2". As it is shown in Figure 2.8 (a), there are more than one possible process. The work produced from a thermal process is:

$$W = \int_{V_1}^{V_2} p dV. \tag{2.26}$$

It can be illustrated by the area surrounded by the process path/curve and the vertical limits $V = V_1$ and $V = V_2$ (shaded area in Figure 2.8 (b))

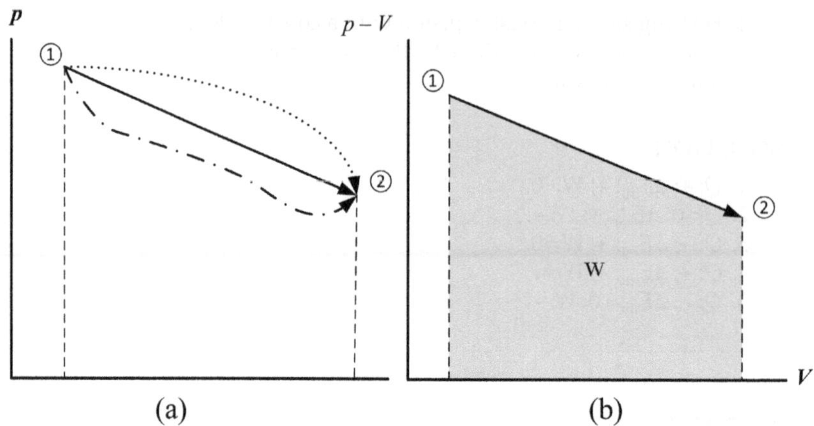

FIGURE 2.8
p-V diagram.

The four basic thermal processes are under the condition of constant temperature, constant volume/density, constant pressure, and no heat transfer, respectively.

Isothermal Process

If the temperature of a gas system is kept constant in a thermal process, this process is called an *isothermal process*. For example, when a balloon is compressed slowly at the atmospheric conditions, the gas inside the balloon would be under isothermal process, while its pressure and volume would change.

When a gas system changes from state 1 to state 2 at isothermal conditions, according to the Ideal Gas Law:

$$p_1V_1 = p_2V_2 \text{ or } \frac{p_1}{\rho_1} = \frac{p_2}{\rho_2}. \tag{2.27}$$

Example 2.8

The initial pressure of an ideal gas $p_1 = 100000$ Pa, and density $\rho_1 = 1.2$ kg/m^3. After it undergoes an isothermal process, its pressure reaches 200000 Pa, what is its density (ρ_2) now?

SOLUTION

Use Equation (2.27): $\rho_2 = \frac{p_2\rho_1}{p_1} = \frac{200000 \times 1.2}{100000} = 2.4 \text{ (kg/m}^3\text{)}$.

The density after the isothermal process is 2.4 kg/m^3.

Examining the First Law of Thermodynamics (2.25), we can see that that the heat energy received by this system is converted into the work to outside: $dQ = 0 + dW$. Because the temperature is unchanged, the change of the internal energy would be "0".

Isobaric Process

If the pressure of a gas system is kept constant in a thermal process, this process is called an *isobaric process*. For example, gas expanding inside a cylinder-piston system to push the piston carrying with a constant load at a constant speed. In this case, its volume and temperature may change in the process, but the pressure is unchanged.

When a gas system change from state 1 to state 2 isobarically, according to the Ideal Gas Law:

$$\frac{V_1}{T_1} = \frac{V_2}{T_2} \text{ or } \rho_1 T_1 = \rho_2 T_2. \tag{2.28}$$

Example 2.9

The initial temperature of gas is T_1 = 35°C, and density is ρ_1 = 1.2 kg/m³. After gas undergoes an isobaric compression process, its density reaches 1.5 kg/m³. What is its temperature (T_2) after the compression?

SOLUTION

Use Equation (2.28): $T_2 = \frac{\rho_1 T_1}{\rho_2} = \frac{1.2 \times (273 + 35)}{1.5} = 246.4\ (\text{K})$.

The temperature after the isobaric compression is 246.4 K, or – 26.6°C.

Analyzing the isobaric process by the First Law of Thermodynamics (2.23), in this process, there is heat transfer between the system and outside, and the internal energy change depends on the temperature difference $\Delta E_{\text{int}} = mC_V(T_2 - T_1)$ while the amount of work depends on the change of volume, $W = p(V_2 - V_1)$.

Isochoric (Isovolumetric) Process

If the volume/density of a gas system is kept constant in a thermal process, this process is called an *isochoric, or isovolumetric process.* In this process, pressure changes linearly with the absolute temperature. The mechanical bulbs used as thermometers in traditional aircraft are designed on the basis of the isochoric process: A certain amount of gas is sealed inside a Bourdon tube, whose volume is fixed; the pressure of the gas increases/decreases when the temperature of the gas increases/decreases. The pressure change can be detected by a Bourdon tube, and then signals of the pressure change can be converted to temperature change as output of a mechanical bulb, which was used in the design of early airplanes.

When a gas system changes its state in an isochoric process, according to the Ideal Gas Law:

$$\frac{p_1}{T_1} = \frac{p_2}{T_2} \text{ or } \frac{p_1}{p_2} = \frac{T_1}{T_2} \tag{2.29}$$

Example 2.10

The initial temperature of gas is $T_1 = 35°C$, and pressure is $p_1 = 100000$ Pa. Find the temperature (T_2) when the pressure of the gas increases to 115000 Pa.

SOLUTION

Use Equation (2.29): $T_2 = \frac{p_2 T_1}{p_1} = \frac{115000 \times (273 + 35)}{100000} = 354.2$ (K).

The temperature has increased to 354.2 K, or 81.2°C.

Examining the equation of the First Law of Thermodynamics (2.25), the heat energy transferred to the system all is converted into the internal energy change of the system, and the term of work pdV is "0", because dV is "0" in the isochoric process. So $dQ = dE_{int}$.

Figure 2.9 shows the p-V diagrams of three routes: (a) gas system expands from state "1" to state "2" in an isothermal process; (b) the same start state "1" and final state "2": the system decreases its pressure while the volume/density is constant and then expands while pressure is constant or isobaric; (c) the system expands from state "1" while pressure is kept constant, isobaric, first and then pressure decreases while its volume is kept constant, isochoric, to state "2". The shadow area under each route

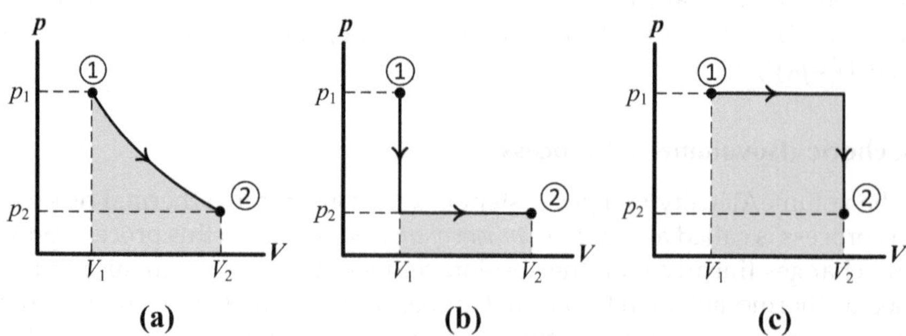

FIGURE 2.9
Work output of different thermal processes.

shows that the work outputs of the system are very different with different processes even if the system starts from the same state and finishes at the same state. Therefore, it is very important to understand these thermal processes.

Adiabatic Process

If there is no heat transfer during a thermal process, $dQ=0$, this process is called an *adiabatic process*. For a reversible adiabatic process, it is also called *isentropic* process. A gas system converts its internal energy to work output according to the First Law of Thermodynamics: $-dE_{int}=dW$, if it is in an adiabatic process. For example, the process of air flowing over an aerofoil is commonly treated as an adiabatic process, because there is no heat exchanged when the air passes the aerofoil. This assumption will be used often in the following chapters.

The gas property changes from one state, "1", to another state, "2", through an adiabatic process assuming at every moment during the process the state of the gas always follows the Ideal Gas Law. Then the relationship of the gas property at state "1" and state "2" is described by the following formulae:

$$p_1V_1^\gamma = p_2V_2^\gamma \quad \text{or} \quad \frac{p_1}{\rho_1^\gamma} = \frac{p_2}{\rho_2^\gamma}$$
$$\frac{p_1^{\gamma-1}}{T_1^\gamma} = \frac{p_2^{\gamma-1}}{T_2^\gamma} \tag{2.30}$$

which is derived from the Ideal Gas Law and the First Law of Thermodynamics. In these formulae, γ is the adiabatic index: $\gamma = \frac{c_p}{c_V} = \frac{C_p}{C_V}$. c_p [J(mol.K)$^{-1}$], and C_p[J(kg.K)$^{-1}$] are specific heats of the fluid when pressure is constant, while c_V [J(mol.K)$^{-1}$], and C_V [J(kg.K)$^{-1}$] are specific heats of the fluid when volume/density is constant. For air, $\gamma = 1.4$.

All of the thermal processes can only occur one at a time. For example, if air in a syringe is compressed under a constant pressure, isobaric, it cannot be an isothermal or adiabatic process. Its temperature has to change in the process, and heat will be released in the process as well.

Example 2.11

The initial temperature of an air system is T_1 = 35°C, and pressure is p_1=100000 Pa. After an adiabatic process, its pressure reduces to 50000 Pa. What is the temperature (T_2) of the system? How much work does this system produce? (C_V=717.5 J(kg.K)$^{-1}$)

SOLUTION

Use formula (2.30) and $\gamma = 1.4$ for air:

$$T_2^{\gamma} = \frac{p_2^{\gamma-1}T_1^{\gamma}}{p_1^{\gamma-1}} = \frac{50000^{0.4} \times (273+35)^{1.4}}{100000^{0.4}}$$

$$T_2 = 308 \times \left(\frac{1}{2}\right)^{\frac{0.4}{1.4}} = 252.7\ (\text{K}).$$

The work W: use the First Law of Thermodynamics,

$$Q = \Delta E_{\text{int}} + W = 0: \frac{W}{m} = -\frac{\Delta E_{\text{int}}}{m} = -C_V(252.7 - 308) \approx 39.7\ \text{kJ.kg}^{-1}$$

This system has done 39.7 kJ of work per kg gas.

Energy Equation

In this section, we use the principle of the First Law of Thermodynamics to analyze a defined gas flow system as shown in Figure 2.1. When the gas flows through this system in a steady state, the states of the gas and the flow rate at the inlet and outlet of this system do not change, and the total mass within this system is constant. This defined system can be treated as a closed system, so the First Law of Thermodynamics, Equation (2.25), can be used for this defined gas flow system.

Equation (2.25) shows the change between the temperature of the system and the work output from the system if there is no heat exchange in or out of this system. For an ideal gas flow system, e.g. as in Figure 2.1, which contains n moles of gas, the equation of the First Law of Thermodynamics (2.25) can be written as:

$$dQ = nc_V dT + pdV \tag{2.31}$$

where n is the number of moles of gas in the system and c_v is the specific heat per mole of gas, when the volume is constant, [J(molK)$^{-1}$].

Differentiate Ideal Gas Law (2.20):

$$Vdp + pdV = nRdT; \text{ then } pdV = nRdT - Vdp \tag{2.32}$$

Substitute (2.32) into (2.31):

$$dQ = nc_V dT + nRdT - Vdp = n(c_v + R)dT - Vdp \tag{2.33}$$

Then (2.33) can be rewritten:

$$dQ = m\left[\frac{(c_V + R)}{M}dT - \frac{V}{m}dp\right] = m\left(C_p dT - \frac{1}{\rho}dp\right) \tag{2.34}$$

where $c_V + R = c_p$, (Serway and Jewett, 2008) and $C_V + R_M = C_p$; M is the molar mass of the gas; and $C_p = c_p/M$ is the specific heat in $J(kg.K)^{-1}$ when pressure is constant.

Consider a steady flow of gas along a level streamline, according to the Euler equation (2.8):

$$\frac{1}{\rho}\frac{dp}{ds} = -v\frac{dv}{ds}, \text{ so } -\frac{1}{\rho}dp = vdv. \tag{2.35}$$

Substitute (2.35) into (2.34):

$$dQ = m(C_p dT + vdv) \tag{2.36}$$

If there is no heat exchange externally, the gas system is in an adiabatic process, $dQ = 0$, i.e.:

$$C_p dT + vdv = 0 \tag{2.37}$$

Integrate (2.37):

$$C_p T + \frac{v^2}{2} = h_t + \frac{v^2}{2} = E \tag{2.38}$$

where E is a constant, $[Jkg^{-1}]$; $C_pT = h_t$ is called (specific) *Enthalpy – thermodynamic potential* $[Jkg^{-1}]$. Equation (2.38) is called the *Energy equation*, which states that *the sum of thermodynamic potential energy and kinetic energy of gas of unit mass along a level streamline is constant, if this isolated system has no heat exchange externally (adiabatic).* The gas system described by the Energy equation is not incompressible, since the equation is derived under condition of adiabatic process. The flow speed of the gas system is not under any limit. It can be low subsonic or high supersonic.

The Energy equation (2.38) shows that the thermodynamic potential will increases if the speed of a gas flow decreases, and vice versa if there is no heat exchange in the process. The increase of thermodynamic potential means the increase in the temperature of the gas flow system. When the speed of a steady gas flow increases, and its temperature will decrease, if there is no heat transfer in and out the gas flow. For example, when an aircraft travels at a free-stream speed, v_{fs}, in air with a temperature, T_{fs}, the airspeed increases over its wings to a local speed v_L, and $v_L > v_{fs}$, and the temperature at this location T_L, will be less than that in free-stream: $T_L < T_{fs}$.

Stagnation Temperature

The Energy equation indicates that when the speed of the flow decreases, the temperature of the gas flow increases. So the temperature reaches its highest value when the flow speed is reduced to "0". This maximum temperature possible in the system is called *stagnation temperature.* It occurs at leading edge of an aircraft, for example, the nose of an aircraft.

The stagnation temperature can be calculated by the Energy equation (2.38), and it shows that the higher temperature at the nose of an aircraft, or at the leading edge of wing, the higher the speed at which the aircraft travels. The leading edge becomes white-hot when a space shuttle reenters the Earth's atmosphere at a very high speed. If T_o is the stagnation temperature, it can be calculated by Equation (2.38):

$$C_pT+\frac{v^2}{2}=C_pT_o \text{ then } T_o=T+\frac{\frac{1}{2}v^2}{C_p} \tag{2.39}$$

Because $C_V+R_M=C_p$ and $\gamma=\frac{C_p}{C_V}$, C_p and C_V can be expressed as:

$$C_p=\frac{\gamma R_M}{\gamma-1}. \tag{2.40}$$

$$C_V=\frac{R_M}{\gamma-1}. \tag{2.41}$$

Example 2.12

An aircraft travels at 200 kt under sea level conditions. (a) What is the temperature over the aerofoil where the air speed has increased to 300 kt? (b) Find the stagnation temperature at the leading edge of the aircraft. ($C_p=1004.5$ J/kg.K)

SOLUTION

(a) Air speed 200 (kt) = 200 × 0.514 = 102.8 (ms^{-1}), 300 (kt) = 300 × 0.514 = 154.3 (ms^{-1}), and the sea level temperature $T=288$ K;
Use Equation (2.38):

$$C_pT_1+\frac{v_1^2}{2}=C_pT_2+\frac{v_2^2}{2}$$

$$C_p\times 288+\frac{1}{2}\times 102.8^2=C_p\times T_2+\frac{1}{2}\times 154.3^2$$

$$T_2=288+\frac{102.8^2-154.3^2}{2\times 1004.5}=281.4\text{ (K)}$$

The temperature is 281.4 K or 8.4°C, at the location where airspeed increases to 300 kt.

(b) At the leading edge of the aircraft, the relative local airspeed is "0" – stagnation.

Use Equation (2.38): $T_o = T + \dfrac{\frac{1}{2}v^2}{C_p} = 288 + \dfrac{102.8^2}{2 \times 1004.5} = 293.3\ (\text{K})$

The stagnation temperature is 293.3 K, or 20.3°C.

Exercises

1. A water hose is connected to a water nozzle. The inner diameter of the hose inlet is 10 mm, and the diameter of the nozzle outlet is 2.5 mm. The water flow rate is 0.5 kgs^{-1}. What is the water speed at the inlet of the hose and at the outlet of the nozzle? (Water density is ρ_w is 1000 kg.m^{-3}.)
2. A level airflow, along the streamline, $v_1 = 45$ m/s, $p_1 = 1.03 \times 10^5$ Pa, $p_{d1} = 1250$ Pa, and $v_2 = 32$ m/s. Find ρ_1 T_1, and p_2.
3. A flowrate of lubricant oil is 5 kg/s passing through a Venturi tube. The ratio of areas of the Venturi tube is 0.005 m^2/0.002 m^2. The density of the oil is 850 kgm^{-3}. Calculate the difference of static pressure/head measured by the Venturi.
4. Air flows at 60 m/s at the inlet of a tunnel, where $A_1 = 0.5$ m^2 under sea level conditions. At the exit of the tunnel, the area $A_2 = 0.2$ m^2.
   a. What is airspeed, and air pressure at the exit, if we treat the air as incompressible?
   b. What is the air temperature T_2 at the exit, if it is incompressible?
   c. Recalculate the airspeed v_2 at the exit, assuming that there is no heat transfer within the tunnel if the air temperature at the exit $T_2 = 278$ K. What is the air density at the exit ρ_2?
   d. What is the stagnation temperature and pressure, when this airflow (no heat transfer) encounters the leading edge of an object?

3

Viscous Flow and Boundary Layer

In Chapter 2, gas was treated as ideal gas. An ideal gas is the gas, within which there is sufficient space between the gas molecules, so that the interaction between molecules can be ignored. In reality, the interaction between the molecules of any fluid not only exists, but can be very significant as well. So the real fluids, either liquids, or gases, are viscous fluids. The dynamics of viscous flow is very different from the ideal fluid flow. Viscous fluid flow can produce more drag, in particular, on the surfaces and at the places, where there is a sudden change on the surface of an object and within the viscous flow. For example, on the surface of an aerofoil and the fuselage of aircraft in flight; inside the pipes of hydraulic fluid and lubricate oil; and around the undercarriage when it is down.

This chapter will explain the property of viscous fluid, the characteristics of viscous flow, and the types of drag caused by viscous fluid flow.

Viscosity

Consider a fluid particle in a 2-D fluid flow field, shown in Figure 3.1: u is the velocity of the fluid along its streamline, y-direction is perpendicular to the direction of u. The fluid is assumed as a continuum, a "Newtonian fluid". The force due to the interaction between the fluid particles is shown as τ, the shear stress. τ is proportional to the velocity gradient (change) in y-direction. In Figure 3.1, the direction of the fluid velocity is x, and the direction perpendicular to the velocity is y, then the shear stress can be written as:

$$\tau = \mu \frac{\partial u}{\partial y}; \quad \text{SI unit}: [Nm^{-2}] \tag{3.1}$$

where the proportionality μ is called the *dynamic viscosity of the fluid*; u is the speed of fluid particle; and y is the direction, which is perpendicular to u, as suggested in the reference (Shandong Engineering College, 1979). In fact, τ is the viscous friction on a unit area.

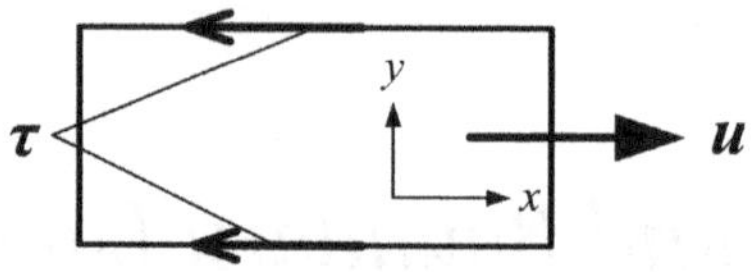

FIGURE 3.1
A particle of viscous fluid.

Viscosity is a property of a fluid. There are two viscosities:

Dynamic viscosity μ: its SI unit: Pa.s, for example, at 20°C, the dynamic viscosity of water is 0.001 Pa.s; and at 20°C, and 1atmosperic pressure, the dynamic viscosity of air is 1.827×10^{-5} Pa.s.

Kinematic viscosity: $\upsilon = \mu\rho^{-1}$, SI unit: m^2s^{-1}.

Effect of Pressure and Temperature on Viscosity

The viscosity of a fluid usually changes with temperature and pressure. For example, honey stored in a cold place appears to be very thick, but when it is heated, it will become runny. This also is true for lubrication oil, hydraulic fluids, and water. Normally, the viscosity of a liquid decreases with the increase of temperature. That is because the bond between liquid molecules becomes more relaxed when its temperature increases, i.e. the molecules are less restricted, and are able to move more freely. Figure 3.2 shows the changes of dynamic viscosity of water with temperature (Figure 3.2 diagram made from the data in Chemical Engineers' Handbook, edited by Perry, 1941).

On the other hand, the viscosity of a gas, for example air or steam, will increase when the temperature increases, as shown in Figure 3.3, and it increases with pressure as well, as shown in Figure 3.4. Figure 3.3 shows the changes of dynamic viscosity and kinematic viscosity of air with temperature at one atmospheric pressure condition. Figure 3.4 shows that the dynamic viscosity of steam changes with temperature at different pressures. The curves in Figures 3.3 and 3.4 are produced from the data in a handbook of thermodynamic and transport properties of fluids (Mayhew and Roger, 1968). The internal energy of the gas increases with its temperature, so molecules of warmer gas move faster than those at a lower temperature. As a result, collisions between the molecules occur more frequently, which, in turn, causes more resistance to the change of relative velocity between them, so the viscosity of the gas increases when its temperature increases. For a similar reason, in general, as pressure increases, the number of collisions between gas molecules increases, which leads to an increase of its viscosity with pressure.

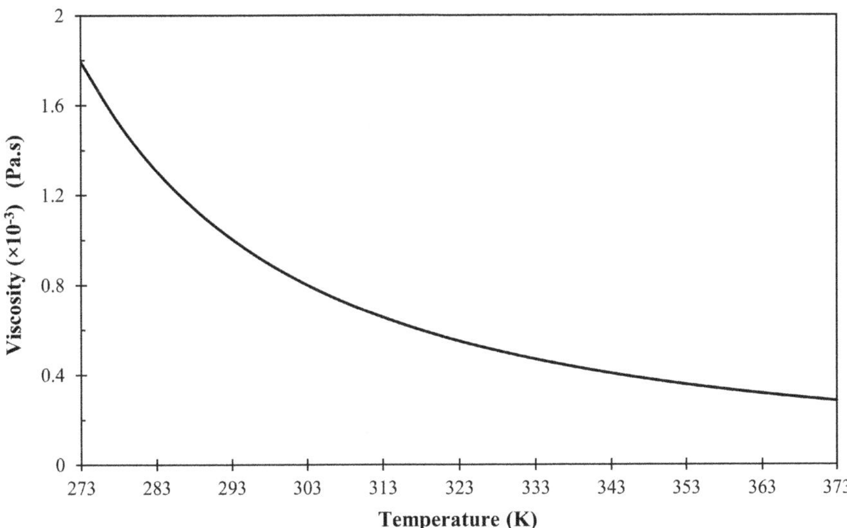

FIGURE 3.2
Change of water viscosity with temperature.

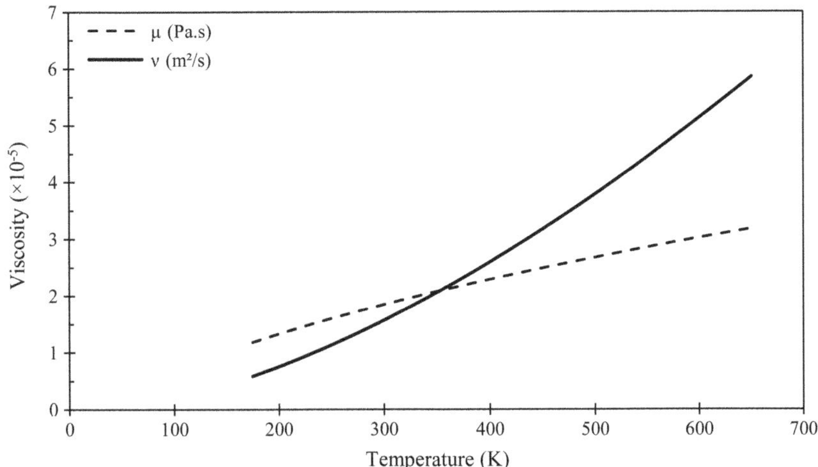

FIGURE 3.3
The changes of dynamic viscosity and kinematic viscosity of air with temperature when pressure is constant.

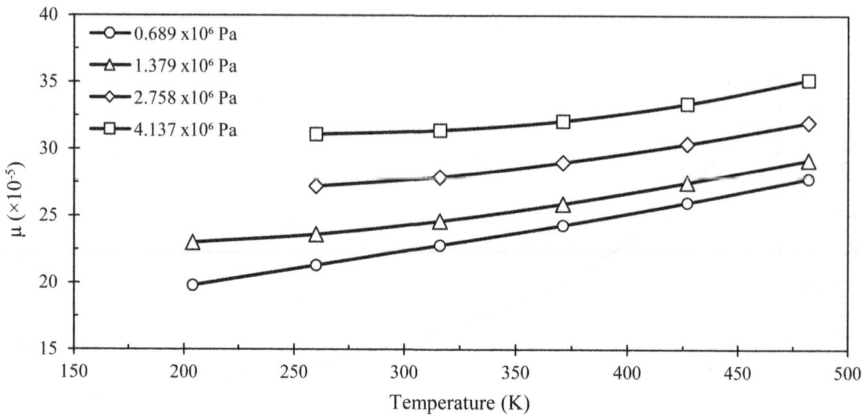

FIGURE 3.4
Changes of viscosity of steam with pressure and temperature.

Pressure Loss

Bernoulli's equation (2.12 or 2.13) is for an ideal fluid, a nonviscous fluid, along a streamline without friction. In reality, all fluids are viscous. Examine an incompressible viscous fluid flow along a streamline as shown in Figure 3.5: the inlet of the flow path is marked section "1" and the outlet section is marked "2". If the fluid were nonviscous, the pressure and velocity of the fluid at the inlet "1" and the outlet "2" would satisfy Bernoulli's equation:

$$p_1 + h_1\rho g + \frac{1}{2}\rho v_1^2 = p_2 + h_2\rho g + \frac{1}{2}\rho v_2^2 = P, \text{ (in Pa)} \tag{3.2}$$

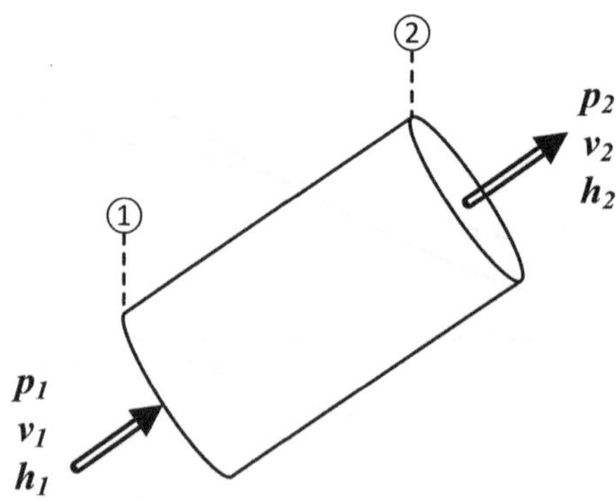

FIGURE 3.5
A piece of viscous fluid flow with the fluid flow properties in two sections.

or:

$$\frac{p_1}{\rho g} + h_1 + \frac{1}{2g}v_1^2 = \frac{p_2}{\rho g} + h_2 + \frac{1}{2g}v_2^2 = H \text{ (in m)} \tag{3.3}$$

When a viscous fluid travels from section "1" to section "2", part of its energy will be consumed to overcome the friction between the fluid particles, and between fluid and the surface of the flow path. So when the fluid reaches section "2", its sum of $\frac{p}{\rho g} + h + \frac{1}{2g}v^2$ will be less than that of at section "1", which is equal to "*H*" because there is a loss of energy due to viscous friction. This loss can be expressed as *loss of pressure head*, h_f, from section "1" to section "2", as shown in Figure 3.3. Equation (3.3) can be rewritten as:

$$\frac{p_1}{\rho g} + h_1 + \frac{1}{2g}v_1^2 = \frac{p_2}{\rho g} + h_2 + \frac{1}{2g}v_2^2 + h_f = H \tag{3.4}$$

The value of h_f depends on the locations of section "1" and "2" as well as the viscosity of the fluid, and the physical size/shape of the flow path. The longer the distance between these two sections, the higher the value of the loss of pressure (head), h_f.

Example 3.1

Water flows as shown in Figure 3.3. The static pressure at both inlet and outlet is atmospheric pressure: 1.013×10^5 Pa, and water density is 1000 kgm^{-3}. At the inlet, the water flows at the speed $v_1 = 8$ ms^{-1}. The difference in height between inlet and outlet is 1 m, and the loss of pressure head for the water flow from inlet to outlet is 0.8 m. What is the water speed at the outlet, v_2? What would be v_2, if the water flow is considered to be a nonviscous flow?

SOLUTION

$p_1 = p_2 = 1.013 \times 10^5$ Pa, because the static pressure at both inlet and outlet is the same.

Use Equation (3.3):

$$\frac{p_1}{\rho g} + h_1 + \frac{1}{2g}v_1^2 = \frac{p_2}{\rho g} + h_2 + \frac{1}{2g}v_2^2 + h_f, \Rightarrow h_1 + \frac{1}{2g}v_1^2 = h_2 + \frac{1}{2g}v_2^2 + h_f$$

So, $\frac{1}{2g}v_1^2 - \frac{1}{2g}v_2^2 = h_2 - h_1 + h_f, \quad \frac{1}{2 \times 9.8}(8^2 - v_2^2) = 1 + 0.8 = 1.8$

$$v_2^2 = 64 - 1.8 \times 2 \times 9.8 = 28.72$$

$$v_2 = 5.36 \text{ ms}^{-1}$$

The speed of water at the outlet is 5.36 ms^{-1}.

If the water is conserved as nonviscous, $h_f = 0$ m.

Then $\frac{1}{2g}v_1^2 - \frac{1}{2g}v_2^2 = h_2 - h_1$, $\frac{1}{2\times 9.8}(8^2 - v_2^2) = 1$,

$$v_2^2 = 64 - 1.0\times 2\times 9.8 = 28.72$$

$$v_2 = 6.66\ \text{ms}^{-1}$$

The water speed would be 6.66 ms^{-1}, if water were nonviscous.

Reynolds Number and Regimes of Viscous Flow

The motion of a nonviscous (incompressible) fluid particle in a one-dimensional steady flow can be described by Equation (2.8). The change of a movement in the direction perpendicular to the movement affects the viscous friction, and the viscous friction causes the change of the movement in the direction of the movement. So the change of motion of a viscous fluid flow has to be at least two-dimensional.

Following the same force-based principles, the change of the velocity of a viscous fluid particle in a two-dimensional steady flow can be described by following equations (Shandong Engineering College, 1979):

$$\begin{aligned} u_x\frac{\partial u_x}{\partial x} + u_y\frac{\partial u_x}{\partial y} &= -\frac{1}{\rho}\frac{\partial p}{\partial x} + \upsilon\left(\frac{\partial^2 u_x}{\partial x^2} + \frac{\partial^2 u_x}{\partial y^2}\right) \\ u_x\frac{\partial u_y}{\partial x} + u_y\frac{\partial u_y}{\partial y} &= -\frac{1}{\rho}\frac{\partial p}{\partial y} + \upsilon\left(\frac{\partial^2 u_y}{\partial x^2} + \frac{\partial^2 u_y}{\partial y^2}\right) \end{aligned} \tag{3.5}$$

where u_x and u_y are velocity components in x- and y-direction, respectively, and υ is the kinematic viscosity of the fluid. The last terms in Equation (3.5) represent the viscous friction exerted on the fluid particle.

Then, commonly, Equation (3.5) can be normalized to generalize the characteristics of the viscous fluid flow, that is to transform each variable from a property with a unit to a dimensionless variable (no unit). For example, use dimensionless velocities: $\tilde{u}_x = \frac{u_x}{v}$, and $\tilde{u}_y = \frac{u_y}{v}$; and dimensionless length: $\tilde{x} = \frac{x}{l}$, and $\tilde{y} = \frac{y}{l}$, where v can be the average speed of the flow, and l can be a characteristic length of the flow field, length of a plate, diameter of a pipe, or chord of an aerofoil.

The normalized Equation (3.5) becomes:

$$\tilde{u}_x \frac{\partial \tilde{u}_x}{\partial \tilde{x}} + \tilde{u}_y \frac{\partial \tilde{u}_x}{\partial \tilde{y}} = -\frac{\partial \tilde{P}}{\partial \tilde{x}} + \frac{\upsilon}{vl}\left(\frac{\partial^2 \tilde{u}_x}{\partial \tilde{x}^2} + \frac{\partial^2 \tilde{u}_x}{\partial \tilde{y}^2}\right)$$
$$\tilde{u}_x \frac{\partial \tilde{u}_y}{\partial \tilde{x}} + \tilde{u}_y \frac{\partial \tilde{u}_y}{\partial \tilde{y}} = -\frac{\partial \tilde{P}}{\partial \tilde{y}} + \frac{\upsilon}{vl}\left(\frac{\partial^2 \tilde{u}_y}{\partial \tilde{x}^2} + \frac{\partial^2 \tilde{u}_y}{\partial \tilde{y}^2}\right) \tag{3.6}$$

In Equation (3.6), every variable is dimensionless. As the result of normalization, there is a dimensionless combination of constants as a factor of the last terms in Equation (3.6): $\frac{\upsilon}{vl} = \frac{\mu}{\rho vl}$, the inverse of which is called the *Reynolds number* $Re = \frac{\rho vl}{\mu}$.

Reynolds Number

Reynolds number, *Re*, is a dimensionless number. For a fluid flow:

$$Re = \frac{\rho vl}{\mu} \tag{3.7}$$

where, ρ is the density of the fluid; v is the flow speed, or the magnitude of the velocity; μ is the dynamic viscosity of the fluid; and l is a characteristic length of the fluid flow. For example, l will be the diameter of a pipe if the fluid flows through a pipe, or a length of a plate, if it flows over a plate.

The Reynolds number can be recognized as a ratio of inertia force of the fluid to the friction force due to the viscosity of the fluid. The Reynolds number can be used to describe the characteristics of the fluid flow. In aerodynamics and fluid dynamics, the Reynolds number is one of the most important "dimensionless numbers", and it determines the regimes of fluid flow.

Regimes of Viscous Flow

Particles of a viscous fluid travel in a flow stream, and their flow pattern can change from orderly to disorderly with the change of flow speed or flow path. Figure 3.6 (a) shows a fluid flow in a pipe: there is a layered flow, and it is getting rippled, and then the layer is broken into small rotating pieces, when the speed of the flow is increased. Figure 3.6 (b) shows the viscous flow around a rod with clear layers when the speed of the flow is low, and with a disordered pattern when the speed of the flow is high. Figure 3.7 shows the change of a flow over flat plate. At the leading edge of the plate, the flow is flat and smooth, and the flow develops into a disorderly unsmooth mixed flow.

So it is clear that a viscous fluid can be in different flow regimes. The viscous fluid flow can be divided into two major regimes: *laminar* flow and *turbulent* flow. For example, the flow shown in the top pipe in Figure 3.6 (a), the

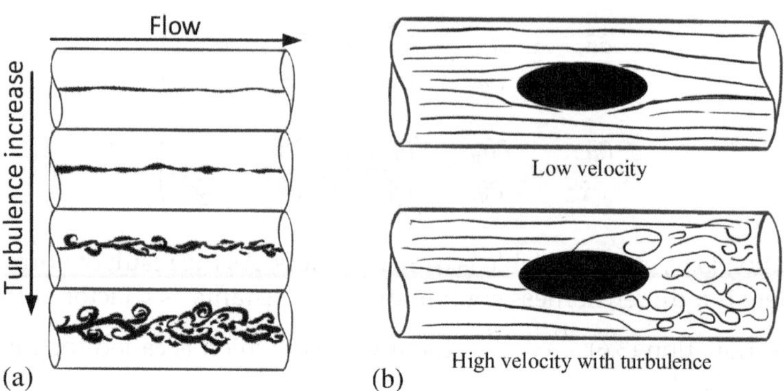

FIGURE 3.6
Schematic diagrams of laminar and turbulent flow: (a) in a pipe; (b) around a cylindrical body.

top picture of Figure 3.6 (b), and the flow over the front part of the plate in Figure 3.5 is laminar flow, in which fluid particles move in an orderly way in layers. The flows shown in Figure 3.6 (a) and over the rear part of the plate in Figure 3.7 are turbulent flow, in which fluid particles may rotate, collide with each other, and move randomly.

The regime of a flow can be determined by a set of Reynolds number rules. These rules were derived from well-controlled experiments and tests. Those rules have worked successfully throughout the development

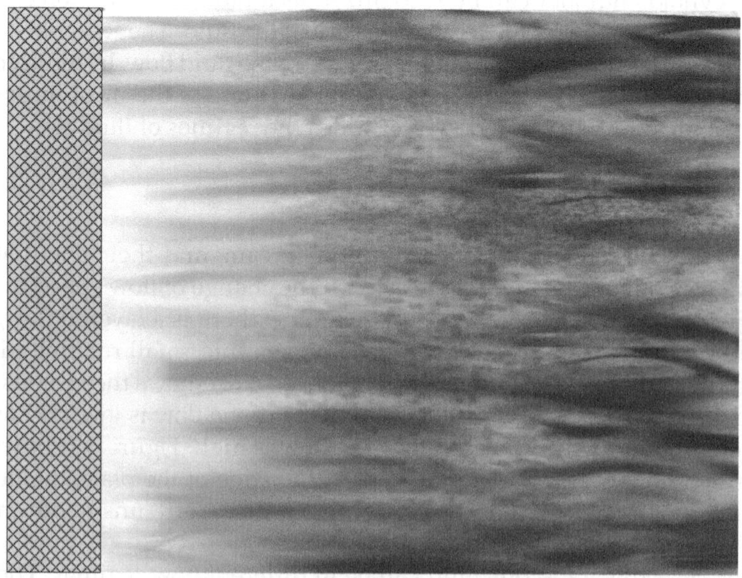

FIGURE 3.7
Flow regime change over a flat plate.

of fluid dynamics/aerodynamics. The set of "rules" may be different for different physical frameworks for fluid flows, and apply no matter whether the fluid is a liquid, or a gas. Here we give an example of Reynolds number rules.

For an internal pipe flow:

The flow is laminar, if its Reynolds number Re < 2000;

The flow is turbulent, if its Reynolds number Re > 4000;

The flow is in transition, if 2000 < Re < 4000.

Example 3.2

Water and air flow through pipes with the same diameter 0.5 m, and both flow at the same speed: 1 ms^{-1}. The densities of water and air are $\rho_w = 1000$ kgm^{-3}, and $\rho_a = 1.225$ kgm^{-3}, respectively. The dynamic viscosities of water and air are $\mu_w = 0.001$ Pa.s, and $\mu_a = 0.181 \times 10^{-4}$ Pa.s, respectively. Calculate the Reynolds number for water and air, and identify the regime of their flow.

SOLUTION

Use the definition of Reynolds number (3.7): $Re = \frac{\rho v l}{\mu}$

For water: $Re = \frac{\rho v l}{\mu} = \frac{1000 \times 1 \times 0.05}{0.001} = 5000$.

It > 4000, and according to the "rules" the water flow is turbulent.

For air: $Re = \frac{\rho v l}{\mu} = \frac{1.225 \times 1 \times 0.05}{0.181 \times 10^{-4}} = 3384$.

2000 < 3384 < 4000, so the airflow is in transition.

Boundary Layers

When a viscous fluid flows pass a surface of an object, like the wall of a pipe, surfaces of aerofoil, and fuselage, the fluid particles, which contact the surface, will "stick" on the surface due to its viscosity, and relative speed of the particles to the surface is "0". So there is a layer of fluid formed near the surface of the object, in which the speed of the fluid particles is smaller than that in the main stream, and gradually increases from 0 to the speed of main stream. This layer is called the *boundary layer.*

Usually a boundary layer is relatively thin comparing with the field of the main stream. It depends on the type of fluid, regime of the flow, and quality of the surface. Figure 3.7 shows a boundary layer flow over a plate.

Structure of the Boundary Layer

With particle tracking and photographic technologies, the structures of a boundary layer have been studied and defined. Figure 3.8 shows the structures of laminar and turbulent boundary layers over a long plate. The descriptions of the structure of the boundary layer and assumptions on fluid speed and pressure within a boundary layer are:

- The speed, u, of a viscous fluid over a horizontal surface of an object, is parallel to the surface shown in Figure 3.8. On the surface, $u=0$;
- Speed of the fluid increases with the distance from the surface in the direction perpendicular to the surface;
- The thickness of a boundary layer, δ, is the distance from the surface to the location where the speed is about 0.99 of the full local speed of the main flow, u_∞, as shown in Figure 3.8;
- The pressure of the fluid in the boundary layer in the direction perpendicular to the surface is *constant*, and the pressure in the direction parallel to the surface will change with the pressure outside the boundary layer;
- Laminar boundary layer – thin, starts from the leading edge of the plate, and the speed of the fluid within the boundary layer increases approximately linearly in the direction perpendicular to the surface;
- Turbulent boundary layer – thicker compared to laminar boundary layer, and the speed increases steeply close to the surface, and then it becomes quite uniform near the edge of the boundary layer;
- There is a transition region between laminar boundary layer and the turbulent boundary layer;

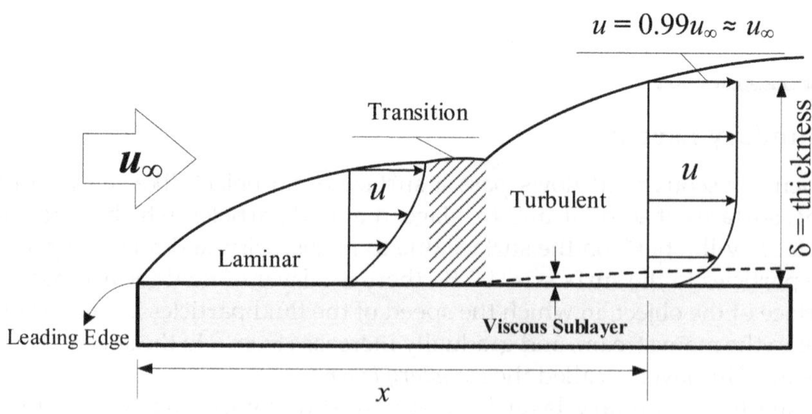

FIGURE 3.8
Structure diagram of a boundary layer.

- The structure of turbulent boundary layer is more complicated: there is a viscous laminar sublayer just over the surface within the turbulent boundary layer, as shown in Figure 3.8;
- When a fluid flows over a long object, the thickness of the boundary layer at the leading edge of the object is "0" and then increases gradually along the surface of the object;
- Initially, the boundary layer close to the leading edge is laminar, then it becomes transitional, and it becomes turbulent after the transition region. This development process of boundary layers can be seen in Figures 3.6 and 3.7;
- The Reynolds number, *Re*, is used to characterize the regime of boundary flow layer.

$$Re_x = \frac{\rho v x}{\mu}$$

where, for a boundary layer flow, x is the distance from the leading edge of the surface to the location concerned.

The rules to determine the regime of fluid flow near the surface of a flat plate are:

The boundary flow is laminar, if its Reynolds number $Re_x < 5 \times 10^5$;

The flow is turbulent, if its Reynolds number $Re_x > 5 \sim 10 \times 10^6$;

The flow is in transition, if $5 \times 10^5 < Re_x < 5 \sim 10 \times 10^6$.

The airflow over an aerofoil can be treated a flow over a flat plate. The "rules" given here are applied to estimate regimes of the boundary flow over an aerofoil for the purposes of this exercise. The high limit of *Re* for flow in transition is a range of values, which depends on many aspects, including the type of fluid, the physical conditions of the fluid, and the condition of the surface – in particular, its roughness. For the purpose of education, 5×10^6 will be used as the upper limit for transitional flow in the example problems and in the exercises that students will carry out.

Example 3.3

Consider airflow over a wing as shown in Figure 3.8 under the conditions at sea level, the wing span is 2.5 m. TAS (true airspeed) of the aircraft is 150 knots. Calculate the local Reynolds number, *Re*, at two locations on the wing, 0.05 m, and 1 m from the leading edge, and identify the regime of the boundary layer flow at these locations. (Assume that the air is at sea level and its dynamic viscosity is $\mu = 1.81 \times 10^{-5}$ Pa.s.)

SOLUTION

At the first location: $x = 0.05$ m, and TAS = 150 kt ≈ 77.1 ms^{-1}.

The Reynolds number: $Re_x = \frac{\rho v x}{\mu} = \frac{1.225 \times 77.1 \times 0.05}{1.81 \times 10^{-5}} \approx 2.61 \times 10^5$.

$Re_x = 2.61 \times 10^5 < 5.0 \times 10^5$, so the flow is laminar.
At the second location: $x = 1$ m, and TAS = 150 kt ≈ 77.1 ms^{-1}.

The Reynolds number: $Re_x = \frac{\rho v x}{\mu} = \frac{1.225 \times 77.1 \times 1.0}{1.81 \times 10^{-5}} \approx 5.22 \times 10^6$.

$Re_x = 5.22 \times 10^6 > 5.0 \times 10^6$, so the flow is turbulent here.

Viscous Flow of Boundary Layer

When a viscous fluid travels over the surface of an object, a boundary layer is formed. There is a change of fluid speed/velocity in the direction perpendicular to the direction of the flow. As described in Equation (3.1), there is viscous shear stress on the fluid particle, when there is a change of velocity in the direction to perpendicular the velocity. The shear stress is in the direction, which is parallel to the velocity, as shown in Figure 3.1. When this shear stress occurs in the boundary layer over an aerofoil, the integration of the viscous shear stress over the surfaces is the total viscous friction to the object, that is, the *skin drag*. So the higher the viscous shear stress near the surfaces, the higher the skin drag is going to be.

The process to obtain the velocity gradient profile within a boundary layer, to determine the skin drag caused by the boundary layer is relatively complex. It requires solutions of the Equations (3.5) or (3.6) together with the differential form of the continuity equation. To solve those equations use either an analytical method with heavily simplified equations or a numerical method with a set of specified assumptions to the initial and boundary conditions (see Shandong Engineering College, 1979; Houghton, 2012; Schlichting, 1978). The knowledge required to solve those differential equations exceeds the scope of the readers of this book; therefore, only basic results will be introduced and explained.

Speed Profile within the Boundary Layer

The solution of the speed distribution within laminar and turbulent boundary layers shown in Figure 3.9 (a) are plotted in Figure 3.9 (b). The plotted curves are the dimensionless boundary layer thickness vs the dimensionless fluid speed, that is $\bar{u} = \frac{u}{u_\infty}$, and $\bar{y} = \frac{y}{\delta}$, where $u_\infty = v$ is free stream speed, or the main stream speed outside boundary layer, and δ is the thickness of boundary layer, which varies along the surface. Figure 3.9 (b) shows that the fluid speed increases approximately linearly within a significant part of laminar

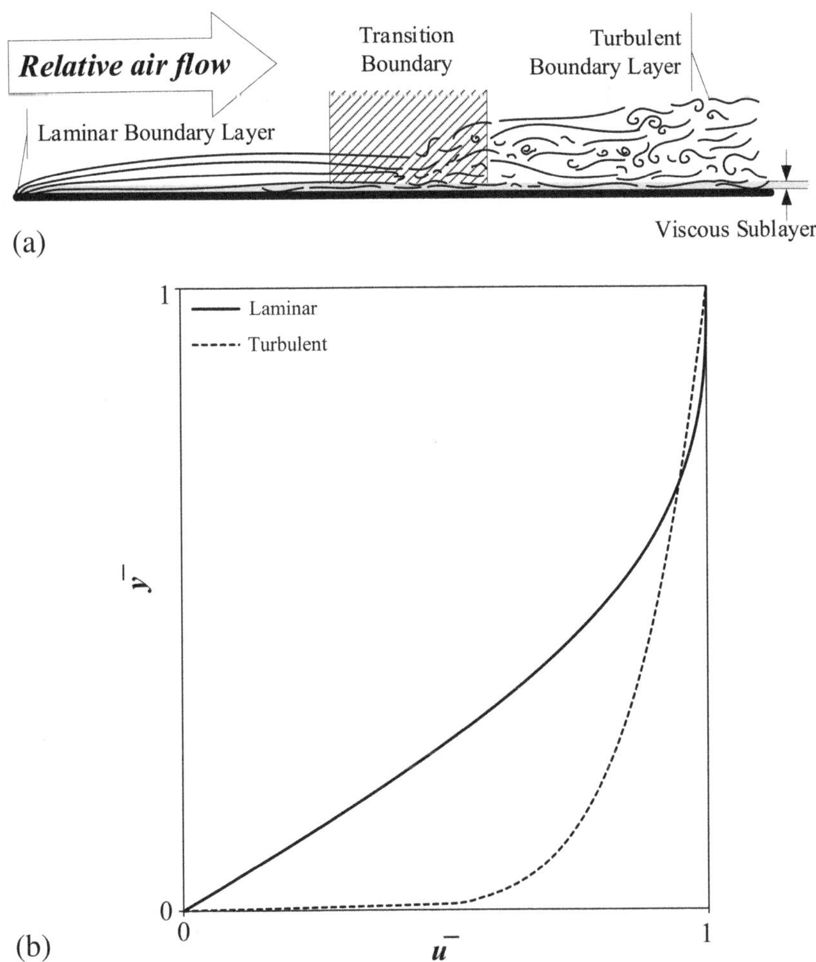

FIGURE 3.9
(a) The development of a boundary layer over a flat plate; (b) the profiles of speed increase along the thickness of laminar and turbulent boundary layers.

boundary layer, while the speed within turbulent boundary layer increases sharply near the surface, reaching approximately 70% of main stream speed within about only 10% of the thickness above the surface.

The dimensionless speed profile labeled "turbulent" in Figure 3.9 (b) is "fuller" than that of laminar layer over the thickness of the boundary layer. Therefore, this indicates that the mean/average speed of turbulent boundary layer is higher than that of the laminar boundary layer, and the kinetic energy in turbulent boundary layer is higher than that of the laminar boundary layer.

Viscous Friction (Skin Drag)

In a viscous level flow, fluid has to consume its kinetic energy to overcome the resistance caused by viscosity of the fluid. As discussed earlier, the speed of fluid within boundary layer is lower than the main stream speed. The longer the fluid travels along the surface, the more kinetic energy will be lost, and the longer distance to take to increase its speed to be the same as it in main stream vertically. So the boundary layer will lose more momentum, and it will get thicker and thicker along the surface.

The expression for viscous friction in a laminar boundary layer is relatively simple, and it is proportional to the speed change in the direction perpendicular to the surface. The expression for viscous friction in a turbulent boundary layer is relatively complex. A laminar sublayer exists next to the surface within the turbulent boundary layer, where the expression for viscous friction for laminar flow can be applied. However, to analyze the dynamic features, for example, velocity change and forces throughout the turbulent boundary layer, a turbulent model is required. This model is a mathematical expression, or a set of equations. Those model equations should indicate that the viscous friction is not in a linear relationship with the change of speed in the direction perpendicular to the surface. There would be more terms of second-order cross derivatives to describe the friction stresses in different directions around a fluid particle, as shown in the terms in the brackets of Equations (3.5) and (3.6). It is difficult to solve those equations for a turbulent boundary layer analytically, and it can be solved numerically with a standard PDE solving program. The details of the turbulent model and the principle of solving the equations of boundary layer can be found in *Boundary-Layer Theory* (Schlichting, 1978).

We can treat the airflow over an aerofoil as the fluid flow over a flat plate. For simple practicable purposes, we adopt the approach described by Shandong Engineering College (1979), which is that viscous friction causes the change of fluid momentum within the boundary layer. So the thickness of a boundary layer at any location along the surface of a flat plate, and the friction force, which is between the fluid and the surface, are estimated and are shown here:

The thickness of the *laminar boundary layer* δ:

$$\delta = 4.64 \times \mathrm{Re}_x^{-0.5} \tag{3.8}$$

where

Re_x = local Reynolds number on the surface/aerofoil, and

x is the distance from the leading edge of plate/aerofoil to a location on the surface.

The friction coefficient C_f on the surface in the *laminar boundary layer*:

$$C_f = \frac{f}{\frac{1}{2}\rho v^2 bc} = 1.372\ \text{Re}_c^{-0.5} \tag{3.9}$$

where $Re_c = \frac{\rho vc}{\mu}$, the overall Reynolds number of the surface/aerofoil; *f* is the viscous friction force; *c* is the length of the surface/chord of the aerofoil; *b* is the width of the surface/the span of the aerofoil; ρ and v are the density and speed, respectively.

The thickness of the *turbulent boundary layer* δ:

$$\delta = 0.383 \times \text{Re}_x^{-0.2} \tag{3.10}$$

where

Re_x = local Reynolds number on the surface/aerofoil, and

x is the distance from the leading edge to the location on the surface.

The friction coefficient C_f on the surface in the *turbulent boundary layer*:

$$C_f = \frac{f}{\frac{1}{2}\rho v^2 bc} = 0.074\ \text{Re}_c^{-0.2} \tag{3.11}$$

where

Re_c = overall Reynolds number of the surface/aerofoil;

f is the viscous friction force;

c is the length of the surface/chord of the aerofoil;

b is the width of the surface/the span of the aerofoil;

ρ and v are the density and speed, respectively.

This friction to the aerofoil's surface is the *skin drag*. So C_f is equivalent to the skin drag coefficient C_{Dskin}.

The skin drag coefficient is the function of the Reynolds number, and the function varies with the regime of the boundary flow as well. Figure 3.10 shows the changing of the skin drag coefficient with flow's Reynolds number *Re*.

The skin drag coefficient decreases with the increase of the Reynolds number. For the same Reynolds number, the skin drag coefficient of the laminar boundary layer is lower than that of the turbulent boundary layer.

The fact that the skin drag near the surface of the turbulent boundary layer is higher than that of the laminar boundary layer can be explained by using the speed profiles in Figure 3.9 (b). According to Equation (3.1), the viscous friction is proportional to the velocity change in the direction, which is perpendicular to the velocity: $\tau = \mu \frac{\partial u}{\partial y}$. In Figure 3.9 (b), it is shown that in the

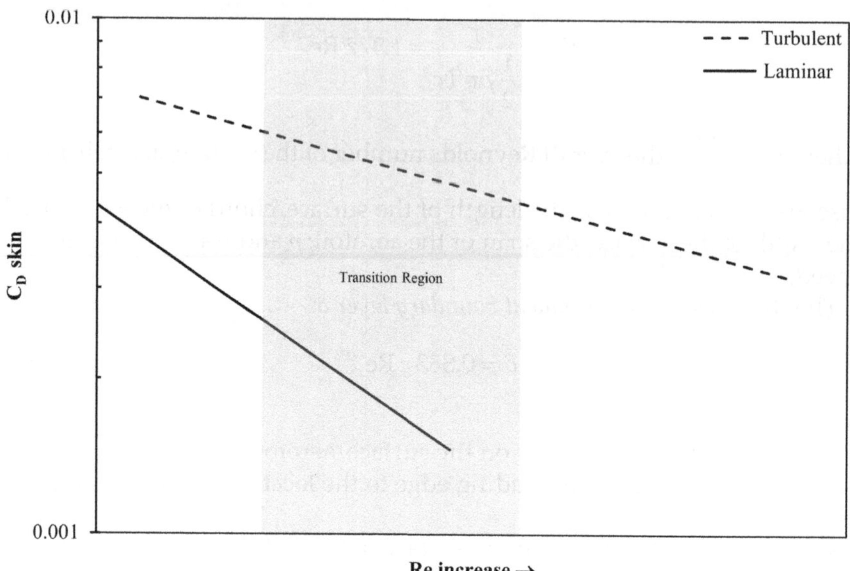

FIGURE 3.10
Skin drag coefficient change with Reynolds number (*Re*).

region near the surface, $\left(\frac{\partial \bar{u}}{\partial \bar{y}}\right)_{\text{turbulent}} > \left(\frac{\partial \bar{u}}{\partial \bar{y}}\right)_{\text{lamnar}}$, so the viscous friction in the turbulent boundary layer could be higher than that in the laminar boundary layer.

The average speed in the turbulent boundary is higher, and it has a higher level of kinetic energy than those of the laminar boundary as we discussed earlier in "Speed Profile in the Boundary Layer". This means the main stream air flow contributes more kinetic energy to a turbulent boundary layer to overcome viscous friction – i.e. skin drag.

Therefore, it is clear from the analysis of the dynamics of the boundary flow that the skin drag of turbulent boundary layer is higher that of skin drag in laminar boundary layer.

Boundary-Layer Separation over a Curved Surface (Stall)

When an incompressible fluid flows over a curved surface, for example, the airflow with the air speed $v<250$ kt over aerofoil of subsonic aircraft, the change of air speed and pressure satisfies the Bernoulli's equation. Figure 3.11 shows the changes of air speed the air pressure in the x–y coordination system over the aerofoil: x-direction – along the surface of the aerofoil, or chord-wise; y-direction – perpendicular to the surface.

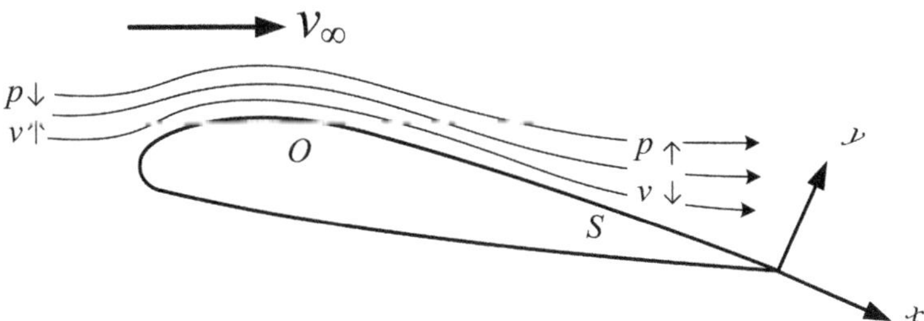

FIGURE 3.11
Viscous air flow over aerofoil in the coordinates of x (along the aerofoil surface) and y (perpendicular to the surface).

According to Bernoulli's equation and the Continuity equation, from the leading edge of aerofoil, the flow path gets narrower, and the local air speed v increases, $\frac{\partial v}{\partial x} > 0$, and the air pressure p decreases, $\frac{\partial p}{\partial x} < 0$.

Within the boundary layer, on the surface: in x-direction: $\frac{\partial p}{\partial x} \neq 0$, p changes chord-wise as the air pressure outside the boundary layer, and $u=0$, where u is the air particle speed in the boundary layer along the surface; in y-direction: $\frac{\partial p}{\partial y} = 0$, p does not change, and air particle speed increases, $\frac{\partial u}{\partial y} > 0$.

When air flows over the most cambered point of the aerofoil, as shown in point O in Figure 3.11, the speed of airflow outside the boundary layer decreases, $\frac{\partial v}{\partial x} < 0$, and air pressure increases, $\frac{\partial p}{\partial x} > 0$. In the boundary layer, on the surface:

in x-direction: $\frac{\partial p}{\partial x} > 0$, $u=0$. When $\frac{\partial p}{\partial x} > 0$, it is called *adverse pressure gradient.*

in y-direction: $\frac{\partial p}{\partial y} = 0$, unchanged, and $\frac{\partial u}{\partial y} > 0$, or $\frac{\partial u}{\partial y} = 0$, or even $\frac{\partial u}{\partial y} < 0$.

Along the surface, there is the viscous friction – skin drag. The speed of the airflow on the surface is "0", but it should be increasing in y-direction, $\frac{\partial u}{\partial y} > 0$. Due to the skin drag, the airflow within the boundary layer loses its kinetic energy, the increase of speed in y-direction gets small and smaller. At some point, there is not enough energy to support the increase of speed: $\frac{\partial u}{\partial y} = 0$, which is the *stagnation point in the boundary layer,* as shown in point "S" in Figure 3.11. In the vicinity of the stagnation point, there are idling (lazy) air particles. Those air particles will be "pushed" away from the surface if it is in adverse pressure gradient, $\frac{\partial p}{\partial x} > 0$, the increase of pressure along the surface (in x-direction). The air particles on the surface leave the surface, move "backward", and form turbulent swirls/eddies. The fact that the air particles leave the surface is called the *boundary-layer separation.*

The boundary layer separation causes a change in the pressure distribution around the aerofoil. The portion of the aerofoil where the boundary layer separates cannot produce lift. So the lift coefficient decreases. The boundary layer separation starts from the trailing edge, and the center of pressure (CP) moves forward when the boundary layer separation takes place. Figure 3.12 (a) and (b) shows the boundary layer separation over an aerofoil. The changes in flow regime of the boundary layer and the increase in the Reynolds number and AoA (Angle of Attack, α) of the aerofoil can cause the separation point moving forward toward the leading edge.

The pressure distribution along the surface of an aerofoil will be changed dramatically when boundary-layer separation occurs too early: close to the leading edge, over the aerofoil, and then the pressure distribution, which produces lift, can be destroyed and the aerofoil loses lift – *stall*. Early boundary separation causes stall.

As discussed earlier, there are two conditions to be met for boundary-layer separation: air particles exhaust the energy to increase speed over the surface, i.e. $\frac{\partial u}{\partial y} = 0$, becoming stagnant; and an adverse pressure gradient at the location, i.e. $\frac{\partial p}{\partial x} > 0$. Then, air particles will move under the higher pressure and leave the surface.

Increasing AoA of a wing can bring the start of boundary-layer separation toward to its leading edge, as shown in Figure 3.12 (b). When AoA increases,

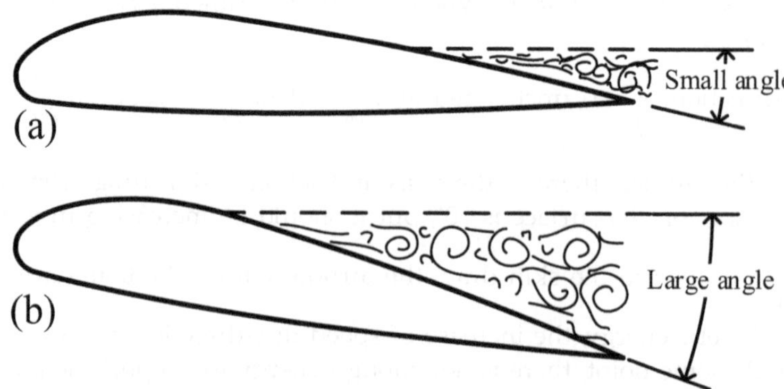

FIGURE 3.12
Boundary-layer separation over aerofoil (a) at a small angle of attack; (b) at a large angle of attack.

it brings the adverse pressure gradient, $\frac{\partial p}{\partial x} > 0$, toward the leading edge, so the boundary-layer separation can happen earlier. The wing will stall if the separation is brought forward too much. The AoA of an aircraft should be closely monitored when it is in take-off – i.e. the climbing phase of flight – to avoid stall. For a subsonic aerofoil, the stall AoA is usually between 15° and 18°.

A laminar boundary layer can separate more relatively easily than a turbulent boundary layer does. Because the average kinetic energy of the air particles within a laminar boundary layer is lower than that in a turbulent boundary layer, the air particle on the surface become stagnant relatively easily, $\frac{\partial u}{\partial y} = 0$. So, under the same adverse pressure gradient condition, the laminar boundary layer can separate more easily.

Boundary-layer separation can occur on control surfaces: an aircraft is in banking and its ailerons deflect, and when the deflecting angle is too large, and it deflects too quickly, the boundary layer over the aileron will separate. In this case, the ailerons cannot produce the correct amount of lift as they are supposed to, and might decrease the lift to cause the aircraft to bank to the opposite direction – *aileron reversal.*

On the other hand, boundary-layer separation can be delayed or prevented if one of, or both conditions can be delayed or eliminated. The methods to reduce or prevent the separation over aerofoil will be discussed in Chapter 4.

Form Drag – Separation Drag

Airflow separation does not only take place over an aerofoil, it also can happen behind different objects, or on other surfaces as well. The shapes of the objects and the quality of the surfaces can cause the passing airflow to leave its streamlining path and form vortices in the wake, as shown in Figure 3.13 (a). Rotating vortices produce a low-pressure zone and cause the fore–aft pressure difference along the object. The airflow around the object loses kinetic energy to form the vortices, and the object needs to overcome this fore–aft pressure difference to move forward. So this fore–aft pressure difference to the object is *form drag*. The form drag is also called *separation drag*. A blunt object or a sudden change of a surface in a fluid flow can disrupt the ability of the flow to follow the contour of the surface and separate, as shown in Figure 3.13 (b).

Vortices can be formed behind an object with different size and intensity. The size and intensity of the vortices change with the *Re* of the flow. Figure 3.14 shows an example of the change of the vortices with the increase of the Reynolds number. Two vortices develop side-by-side behind the cylinder. They will shed away into the wake–flow separation, when it becomes large and strong enough. The vortices will shed one after other alternately

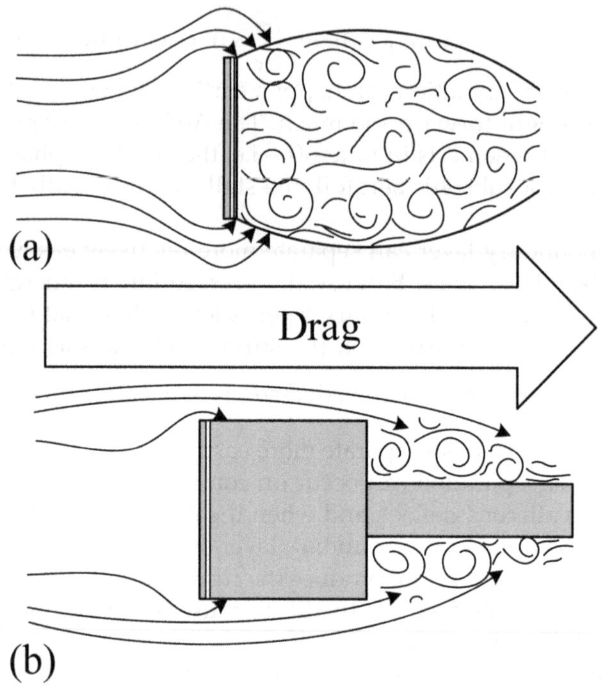

FIGURE 3.13
Fluid flow separation: (a) passing over a vertical plate; (b) passing over a suddenly changing surface.

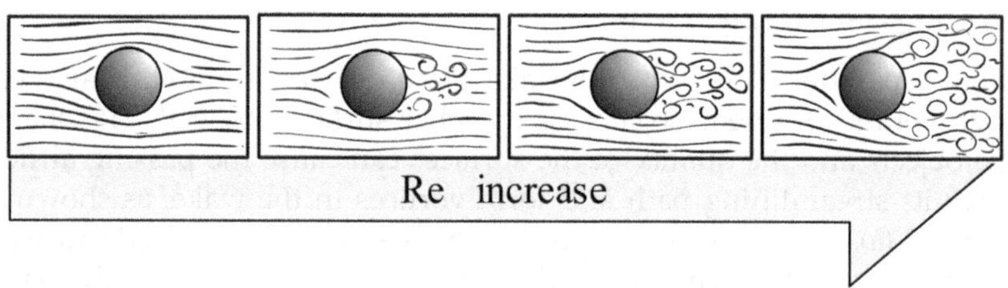

FIGURE 3.14
The development of vortices behind a cylinder with the increase of *Re*.

with a certain frequency. The shedding vortices form a wake behind the cylinder, and this vortex wake is called a *"Karman vortex street"*. The Karman vortex street causes the cylinder to oscillate. A similar oscillation occurs when the boundary layer separates over aerofoil. The frequency f of the Karman vortex street can be determined by Strouhal number, St, as shown here (Schlichting, 1978):

$$St = \frac{fD}{v} \tag{3.12}$$

where
- f is frequency;
- D is the diameter of the cylinder; and
- v is the speed of fluid.

Experiments (Schlichting, 1978) showed that the Strouhal number *St* could increase with the Reynolds number *Re* of the fluid flow. For a fluid flow around a cylinder, when the flow speed increases, *Re* increases, as does

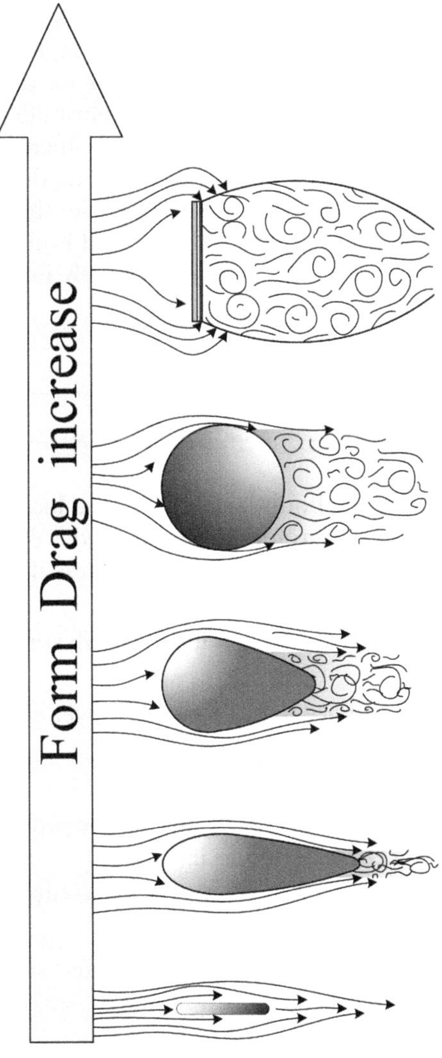

FIGURE 3.15
The comparison of form drag of different objects.

St; therefore, the frequency *f* of vortex shedding has to increase according to Equation (3.12). For example, power lines in wind make noise, because Karman vortex streets are formed around the power lines. The stronger the wind blows, the higher the pitch of noise that the power lines make. Shedding vortices can occur at the downstream side of a chimney, and can also occur behind the strut of fixed undercarriage when the aircraft is in flight. They also occur in boundary-layer separations.

A sudden change in shape – i.e. sudden change in diameter of a cylinder, bumpy surfaces – can affect airflow separation. A streamlined surface design of an object, like an aerofoil, can reduce the airflow separation, and, in turn, can reduce its form drag. Figure 3.15 illustrates the comparison of form drag produced by objects with different surface shapes in the same airflow.

When ice forms on the surface of a wing, the ice changes the profile of the wing surface. It can change the pressure distribution around the wing, increase turbulence in the boundary layer, and increase the thickness of the boundary layer. As a result, the lift of the wing decreases, drag will be increased, and early boundary-layer separation over the wing can take place. This leads to stall, in particular, when the aircraft is climbing. So to prevent ice formation by deicing aircraft wings is extremely important.

Exercises

1. Find the regime of the boundary layer at the location of 0.1 m from the leading edge of an aerofoil, if the airspeed is 80 ms^{-1}, and air condition is at that of sea level. What is the regime if the aircraft travels at the same airspeed at 10,000 ft?
2. Is it correct to say that an object with a smooth surface produces very low form drag?
3. Separation can take place in the boundary layer on a surface or along a flow stream around an object.
   a. What is boundary-layer separation?
   b. Why does the turbulent boundary layer produce a higher skin drag than the laminar boundary layer?
   c. Which boundary layer separates more easily, laminar or turbulent? Why?
   d. Why does boundary-layer separation cause stall?

4

Aerodynamic Forces – Subsonic Flight

Lift L, drag D, thrust T, and weight W are the four fundamental forces on an aircraft in general, as shown in Figure 4.1. The lift and drag are determined by the aerodynamic features of the aircraft, while the aircraft's power plant produces the thrust. Weight is a part of the physical nature of aircraft, which comes from the mass of the aircraft itself and its payload. Lift and drag produced by an aircraft depend on the airflow around the aircraft, in particular, around the aerofoils of the aircraft. Aerofoils include the wings, tail-plane, ailerons, rudder, and other control surfaces of an aircraft. Aerofoils produce "lift" when airflow passes them. The lift makes any maneuver of the aircraft possible, e.g. level flight, banking, climbing/descending, rolling, and yawing. Drag is the resistance to the movement of the aircraft, which is in the same direction as the relative airflow around the aircraft, and it is the most complex force of these four fundamental forces on the aircraft.

Lift, L, and drag, D, can be calculated by the following commonly used formulae, typically stated by PilotBooks (Boys and Wagtendonk, 2011):

$$L = C_L \frac{1}{2} \rho v^2 S \tag{4.1}$$

$$D = C_D \frac{1}{2} \rho v^2 S \tag{4.2}$$

where C_L, and C_D are lift coefficient and drag coefficient respectively; ρ is the local air density, v is the air speed, and $S = bc$ is the area of aerofoil, which is the product of wing span b and chord c. Lift coefficient, C_L, and drag coefficient C_D, are the focus of this chapter. The airflow discussed here is subsonic. The design of the physical appearance of aircraft affects the lift and drag coefficients. The relationship between the coefficients and the features of aerofoil and the shapes of aircraft will be discussed in this chapter as well.

Geometric Features of Aerofoils

The geometric features of aerofoil can affect the nature of the airflow around aerofoil. Some of the features encourage two-dimensional airflow; while some of them enhance three-dimensional airflow. Some of the features

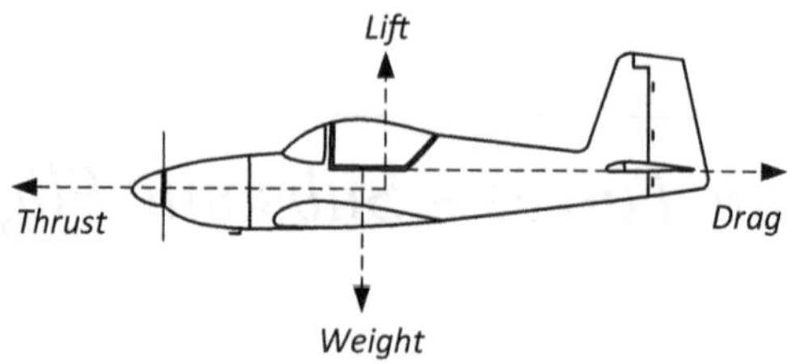

FIGURE 4.1
Main aerodynamic forces on an airplane.

promotes the maintenance of a laminar boundary layer, while others create more turbulence within boundary layer. The airflow around aerofoil with different nature will produce different characteristics of aerodynamics forces, lift, *L*, and drag, *D*. The naming system for aerofoil can provide an easy way to identify the special features of an aerofoil.

Name of Aerofoil

Figure 4.2 shows a schematic diagram of an aerofoil. There are specific terms and lines in Figure 4.2, which highlights the characters of the aerofoil.

1. Leading edge: the forward-most point of the aerofoil (LE);
2. Trailing edge: the rear-most point of the aerofoil (TE);
3. Chord line: the straight line from the leading edge to the trailing edge (the length of the chord line is *c*);
4. Mean camber line: the line marks the middle between the upper surface and the lower surface of aerofoil at any cross section of the aerofoil;
5. Maximum camber: the maximum distance between the mean camber line and the chord line;
6. Maximum thickness: the maximum distance between the upper surface and the lower surface (*t* represents the thickness).

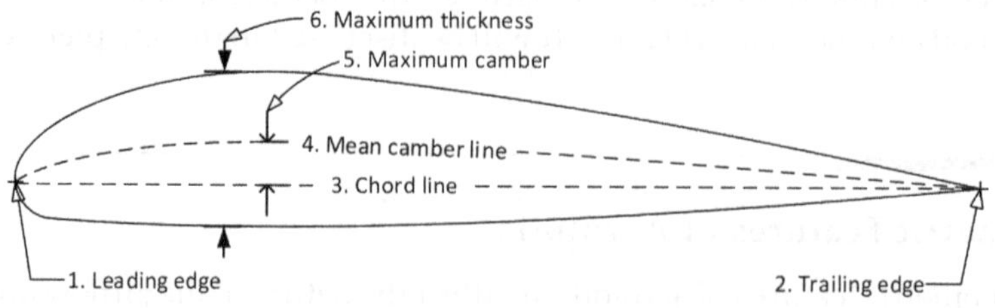

FIGURE 4.2
Geometry of aerofoil.

The best way to describe the geometric feature of an aerofoil is to use standard "language", which is the common rule to provide the measurements of key features of an aerofoil. So aerofoils can be described and compared effectively. NACA's four-digit system is one of the basic ways to name an aerofoil. NACA stands for National Advisory Committee for Aeronautics, which developed this system in early 1930s (Jacobs, Wade, and Pinkerson, 1933). It has been widely used, and five-, six-, or eight-digit systems have been developed since then to suit more complex design and modified design of aerofoil (Marzocca, 2016).

For practical purposes, we introduce the most basic one, the NACA four-digit system: "NACA" followed by four digits/numbers, e.g. NACA4310.

- The first digit is the percentage of the ratio of the maximum camber to the chord length, *c*. In this example, it is "4" – the maximum camber is 4% of chord, *c*.
- The second digit, "3" in the example, indicates the location of the maximum camber from the leading edge in tens of percentage relative to the chord length, *c*: the maximum camber is located at 30% of chord, *c*, from its LE.
- The third and fourth digit indicate the maximum thickness as a percentage of the chord *c*. In this example, the maximum thickness of the aerofoil is 10% of the chord, *c*.

For a symmetrical aerofoil, the chord line and the mean camber line of a symmetric aerofoil are overlapping, so the two digits after NACA will be "00", for example, NACA0010, which means a symmetrical aerofoil with a maximum thickness 10% of its chord *c*. Figure 4.3 (a) shows aerofoil NACA4310, and Figure 4.3 (b) shows aerofoil NACA0010.

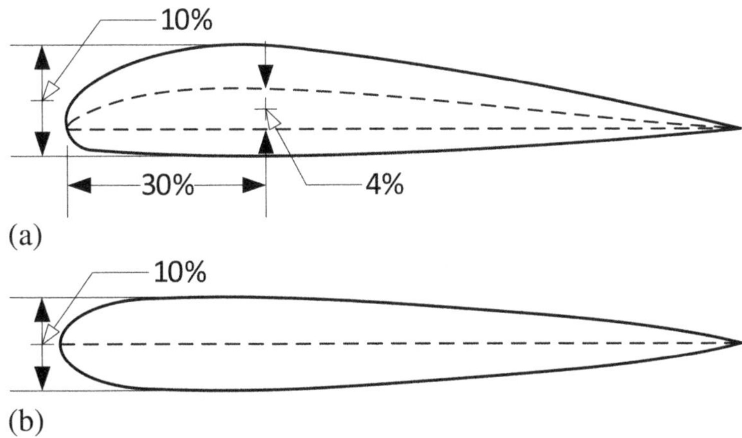

FIGURE 4.3
(a) Aerofoil NACA4310; (b) Aerofoil NACA0010.

Example 4.1

1. There is a high-speed aerofoil with maximum camber of 3% of its chord at 30% of its chord from the leading edge, and its maximum thickness is about 200 mm. The chord of the aerofoil is 2.8 m. Name this aerofoil in the NACA four-digit system.
2. If it is a symmetrical aerofoil with the same thickness, how should it be named?

SOLUTION

1. The maximum thickness in percentage of chord can be calculated as:

$$\frac{0.2}{2.8} = 0.071 \approx 7\%,$$

   So it is NACA3307.
2. For a symmetrical aerofoil, the first two digits should be "0", so it should be NACA0007.

Finite/Infinite Wings

The airflow over a wing can be two-dimensional (2-D), or three-dimensional (3-D). Figure 4.4 (a) shows a 2-D airflow, and the components of air velocity are in two directions: vertical and chordwise. Figure 4.4 (b) shows a 3-D airflow, and the components of air velocity are in three directions: vertical, chordwise, and spanwise.

The length of wing span affects the direction of the air velocity. The spanwise component of the velocity will be significantly reduced if the wing span is very large with regard to its chord, i.e. large aspect ratio. In an extreme situation where the wing is infinitely long, or the wing without wingtip, and

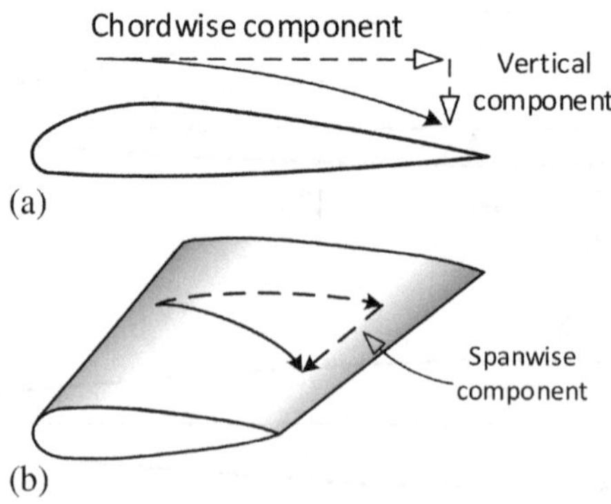

FIGURE 4.4
Components of air velocity over aerofoil: (a) two-dimensional; (b) three-dimensional.

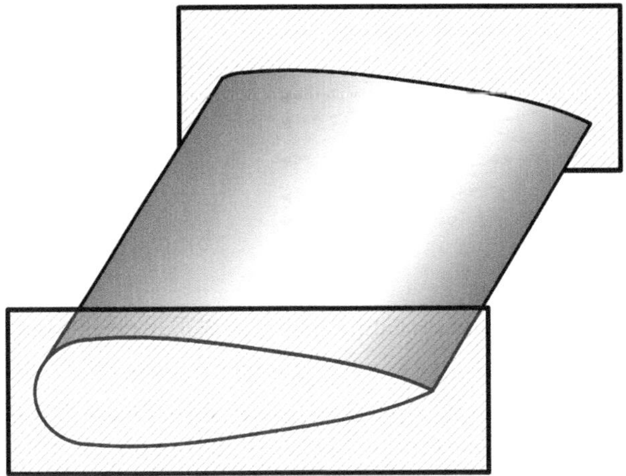

FIGURE 4.5
Schematic diagram of a section of infinite wing used in wind tunnel test.

the airflow over the wing will be two-dimensional. The wing with an infinite wingspan, or without wing tip, over which the air flow is two-dimensional, is called an *infinite wing*. A wing with a very large aspect ratio can be regarded as an infinite wing. An idealized infinite wing can be found as a section of straight wing, which is used for aerofoil development or research in a wind tunnel. This wing does not have a wing root or wingtip as can be seen in the schematic diagram in Figure 4.5.

The wing with limited wing span and wing tip is called a *finite wing*. The airflow over a finite wing is three-dimensional. A finite wing is a real wing. Wings on existing aircraft are generally finite wings.

Theory of Lift

The main function of an aerofoil is to produce lift. The airflow around an aerofoil with a positive angle of attack can be observed with a pattern as shown in Figure 4.6 (a). The streamlines in Figure 4.6 (a) show that airflow speed is higher over the upper surface than the airflow under the lower surface, and there is prominent *downwash* at the trailing edge, and *upwash* at the leading edge. The air pressure around the aerofoil is not uniform as it would be in the free-stream. There is a pressure distribution around the aerofoil, and Figure 4.6 (b) shows the distribution of $\Delta p = p - p_{fs}$ around the aerofoil. The negative sign "–" means that the local pressure is less than the free-stream pressure, and the positive sign "+" means that the local

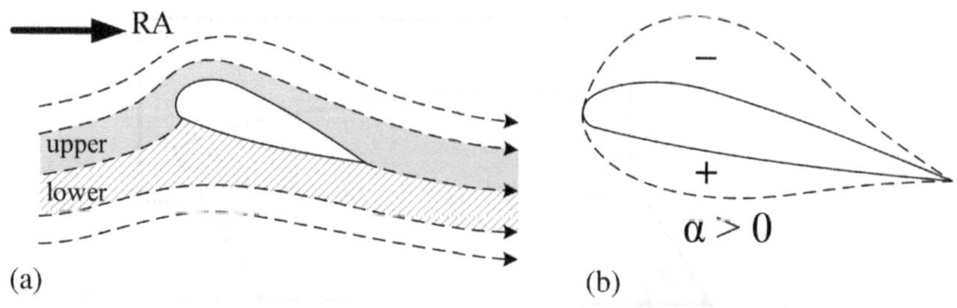

FIGURE 4.6
(a) Streamlines of airflow around aerofoil; (b) Relative pressure distribution around an aerofoil.

pressure is greater than the free-stream pressure. Overall, the pressure on the lower surface of the aerofoil is higher than that on the upper surface, so we know, in principle, the aerofoil produces lift. There are various theories of lift which have been developed to quantify the lift. They all depend on the descriptions or the assumptions about the airflow field around the aerofoil. Some of these are simple and relatively easy to understand, but operate under very strict assumptions, for example, the Bernoulli theorem. Some of them are very comprehensive with detailed mathematical description of the fluid field around aerofoil, for example, 3-D computational simulations of Navier–Stokes equations. However, also, some of them are well-used, for example, in aviation training, with limited description of aerofoil and airflow, and not too complicated to explain the basic relationship between lift and AoA, for example, the Circulation theory of lift.

Bernoulli's Theorem

If airflow is subsonic and air can be treated as incompressible, following the stream lines shown in Figure 4.6 (a), air speed and pressure around the aerofoil can be analyzed by Bernoulli's equation. Assuming it is level flow, because the elevation difference between the free-stream and over the aerofoil is negligible comparing the size of the aircraft. So Bernoulli's equation contains only two terms as shown in Equation (2.14), one-dimensional – along streamline: $p+\frac{\rho}{2}v^2=P$; and the air flow is considered as incompressible, and friction-free fluid flow. The total pressure for every streamline is the same in free-stream before the leading edge. The air speed increases when the flow path between two streamlines narrows; the air speed decreases when the flow path between two streamlines widens. Figure 4.6 (a) shows that the speed of the airflow over the aerofoil increases, while the speed of the airflow under the aerofoil decreases. Airspeed v is a function of streamline s, v (s), along the surface of the aerofoil, and we can obtain the force per wing span on the aerofoil by the integral as following L:

$$\begin{aligned} L &= \int_{\text{Lower Surface}} p ds - \int_{\text{Upper Surface}} p ds \\ &= \int_{\text{Lower Surface}} \left(P - \frac{\rho}{2} v(s)^2\right) ds - \int_{\text{Upper Surface}} \left(P - \frac{\rho}{2} v(s)^2\right) ds \\ &= \int_{\text{Upper Surface}} \frac{\rho}{2} v(s)^2 ds - \int_{\text{Lower Surface}} \frac{\rho}{2} v(s)^2 ds \end{aligned} \tag{4.3}$$

The speed is higher over the upper surface than that over the lower surface, so the result of (4.3) is positive. The aerofoil produces lift. This analysis of lift is direct, simple, and easy to understand. However, the integration in Equation (4.3) requires the assumption on the function $v(s)$ to be able to give a value of lift.

In fact, the pressure distribution around aerofoil changes with its AoA. (If you read any principle of flight for pilots, you know that the pattern of the pressure distribution will change with AoA). From the formula (4.1) for calculating lift, it is quite clear that the lift coefficient C_L in formula (4.1) should be changing with the pressure distribution. In other words, the lift coefficient C_L should change with AoA, and should be a function of AoA. But Bernoulli's equation cannot establish the relationship between the pressure distribution and angle of attack, and it cannot explain the upwash at the leading edge and the downwash at the trailing edge of aerofoil, and the upwash and downwash changes with AoA as well.

PDE Aerodynamics System

The physical and dynamical nature of an airflow field around the aerofoil can be described by a set of partial differential equations (PDE), which are usually derived from the fundamental principles of mass conservation, energy conservation, Newton's motion theory (momentum), and the Ideal Gas Law. A set of PDEs for a particular airflow field can also be called as a mathematical model of this airflow system. The airflow system can be the airflow over a wing; can be the airflow around the whole aircraft; can be just a boundary layer on any part of the surface of the body of aircraft. The airflow can be compressible or incompressible, and it can be one-, two-, or three-dimensional.

For example, for an *incompressible* subsonic airflow over a wing, the air is an ideal gas with properties of pressure p, density ρ, and temperature T. The air velocity vector $\mathbf{v} = u\vec{i} + v\vec{j} + w\vec{k}$, where u, v, and w are the components of air particle velocity in x- (the direction of relative airflow, or chordwise), y- (vertical), and z-direction (spanwise), respectively. Then the following set of PDEs can describe this air flow system:

$$\text{Continuity equation:} \quad \frac{\partial u}{\partial x} + \frac{\partial v}{\partial y} + \frac{\partial w}{\partial z} = 0 \tag{4.4}$$

Motion equation (momentum equation):

$$\rho\left(\frac{\partial u}{\partial t}+u\frac{\partial u}{\partial x}+v\frac{\partial u}{\partial y}+w\frac{\partial u}{\partial z}\right)=\rho g_x-\frac{\partial p}{\partial x}+\frac{\partial \sigma_{xx}}{\partial x}+\frac{\partial \sigma_{xy}}{\partial y}+\frac{\partial \sigma_{xz}}{\partial z}$$

$$\rho\left(\frac{\partial v}{\partial t}+u\frac{\partial v}{\partial x}+v\frac{\partial v}{\partial y}+w\frac{\partial v}{\partial z}\right)=\rho g_y-\frac{\partial p}{\partial y}+\frac{\partial \sigma_{yx}}{\partial x}+\frac{\partial \sigma_{yy}}{\partial y}+\frac{\partial \sigma_{yz}}{\partial z} \qquad (4.5)$$

$$\rho\left(\frac{\partial w}{\partial t}+u\frac{\partial w}{\partial x}+v\frac{\partial w}{\partial y}+w\frac{\partial w}{\partial z}\right)=\rho g_z-\frac{\partial p}{\partial z}+\frac{\partial \sigma_{zx}}{\partial x}+\frac{\partial \sigma_{zy}}{\partial y}+\frac{\partial \sigma_{zz}}{\partial z}$$

Where ρg_x, ρg_y, and ρg_z, are the components of per unit volume body force, includes gravitational force, in *x*-, *y*-, and *z*-direction respectively; $\frac{\partial p}{\partial x}$, $\frac{\partial p}{\partial y}$ and $\frac{\partial p}{\partial z}$ are the pressure gradients in *x*, *y* and *z* respectively; σ is a matrix, shown in Equation (4.6), of surface force on air particles, including three normal forces: diagonal elements in the matrix, and six shear forces: two in each direction):

$$\sigma=\begin{pmatrix}\sigma_{xx} & \sigma_{xy} & \sigma_{xz}\\ \sigma_{yx} & \sigma_{yy} & \sigma_{yz}\\ \sigma_{zx} & \sigma_{yy} & \sigma_{zz}\end{pmatrix} \qquad (4.6)$$

The shear forces are viscous friction, the form of which can vary depending the regime of the flow, laminar or turbulent flow. The shape of the wing is part of the boundary of this airflow system. Additionally, it needs to determine the initial condition, such as, if it is steady state, what the air properties and velocity are when time is "0"; and the boundary conditions of the airflow system such as, on the surface of the wing air velocity is "0". Then the mathematical model is complete for the airflow over a wing. The solutions of the set of PDEs will shows the velocity of air particles, and the air pressure distribution within this airflow system, and then the lift, and drag produced on the wing can be estimated.

To solve the set of PDEs analytically with its specified initial condition and boundary conditions is difficult, and in many cases of flow fields it is not possible. This partial differential equation system currently can be solved numerically. There are many commercially available software packages for CFD (Computational Fluid Dynamics). The process of solving the PDEs and displaying the values of results is mathematical simulation. Making assumptions on boundary layers, and initial and boundary conditions, CFD provides approximate solutions of fluid field. CFD simulation is a very powerful tool, and it is widely used by academics, and aircraft designers to analyze the aerodynamic forces/features of an aircraft, to evaluate a new design, and to investigate problems of existing design. Figure 4.7 shows one of examples of

FIGURE 4.7
Example of CFD numerical simulations of airflow under aircraft (Credit: NASA).

CFD simulations: turbulent vortices and their interference under an aircraft when its landing gear is down.

However, to use this method requires a sound knowledge in applied mathematics and sophisticated training in computing techniques apart from solid understanding in aerodynamics. For professional pilots, this method of analyzing lift is not very convenient.

Circulation Theory of Lift

The Circulation theory of lift explains the upwash at leading edge and downwash at the trailing edge of the airflow around aerofoil, and can establish/confirm the linear relationship between lift coefficient and AoA, and the airflow around aerofoil is assumed as two-dimensional flow (i.e. over an infinite wing). This method of analyzing lift is easy to understand, and the results, in general, agree with observations in real life, even though there are certain limitations and approximations to the aerofoil and airflow in the theory. It has been widely used by pilots, technicians and engineers in the aviation industry.

From the experimental observations of an aerofoil in an air stream reported by Prandtl and Tietjens (1957) and quoted by Panaras (2012), it was clear that there is an upwash at the leading edge of an aerofoil, and there is a downwash at the trailing edge. The upwash and downwash increase with the increase of AoA, and the downwash is stronger than the upwash.

Figure 4.8 (a) and (b) show the schematic diagrams to show the 2-D airflow passing a very thin cambered aerofoil (maximum camber is at 50% of *c*), while the schematic diagrams in Figure 4.8 (c) and (d) shows the 2-D airflow passing a very thin symmetrical aerofoil. The thin aerofoil sits in the airflow at AoA = 0 in Figure 4.8 (a), and (c), and the thin aerofloil sits in the airflow at a small positive AoA in Figure 4.8 (b), and (d). The both of upwash and downwash are greater when the AoA is greater. The streamlines deflect with a greater degree when AoA > 0, than that when AoA = 0. The magnitude of vertical component of the velocity in the upwash is less

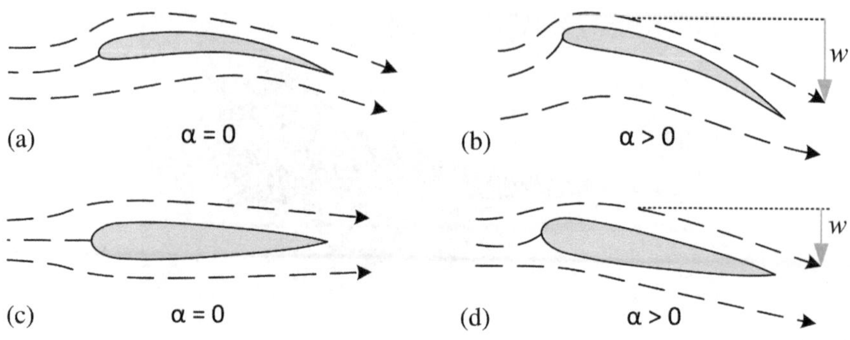

FIGURE 4.8
Streamlines of airflow around aerofoils: (a) a thin cambered aerofoil at "0" AoA; (b) a thin cambered aerofoil at a small positive AoA; (c) a thin symmetrical aerofoil at "0" AoA; (d) a thin symmetrical aerofoil at a small positive AoA.

than that in the downwash, when the AoA is positive shown in Figure 4.8 (b) and (d), where v is relative airspeed, and w represents downwash. These schematic diagrams also show that the local air speed over the aerofoil is greater than that at the free-stream, while the local air speed under that aerofoil is lower than that at free-stream except the symmetrical aerofoil at AoA = 0 in Figure 4.8 (c).

Examine the airflow around the thin symmetrical aerofoil: the local air velocity has changes from the free-stream velocity either in magnitude, or in direction, or both. Comparing the free-stream velocity v, which is horizontal and uniform, the local air velocity in the upwash at the leading edge is the sum of free-stream air velocity v and a small vertical velocity w^+; the local air velocity over the upper surface of the aerofoil is the sum of free-stream air velocity v and a small horizontal velocity $w_{+;}$ the local air velocity in the downwash at the trailing edge is the sum of free-stream air velocity v and a small vertical velocity w^-; and the local air velocity under the lower surface of the aerofoil is the sum of free-stream velocity v and a small horizontal velocity w_-, as shown in Figure 4.9. From this diagram, it is reasonable to suggest that the airflow around the aerofoil is a superposition of two simple airflows: a uniform airflow with a velocity v, and a circulation, which follows $w^+\uparrow$, $w_+\rightarrow$, $w^-\downarrow$, and $w_-\leftarrow$, around the aerofoil. We know in fact that the magnitude

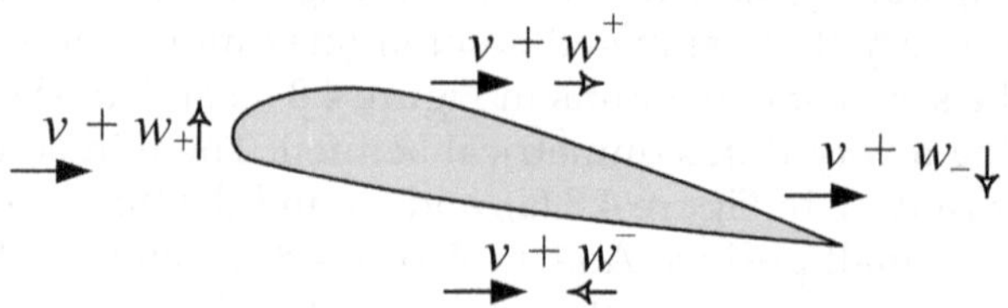

FIGURE 4.9
Illustration of the assumption on air velocity around aerofoil.

of the speed of the circulation around aerofoil is not uniform, e.g. $|w^-| > |w^+|$, and the circulation is induced by the downwash, and it will intensify if the downwash increases. So in order to capture the essence of the nature behavior and simplify the analysis, we make an approximation that the air speed in the circulation is constant, w, in all the directions, and its magnitude is less that of the downwash: $|w| < |w^-|$, but is increases linearly with the magnitude of the downwash, $|w| \propto |w^-|$. The airflow around an aerofoil can be displayed as the superposition shown in Figure 4.10.

In early 1900s, Kutta and Joukowski proposed a theory that stated that there was a force generated when a circulation exists in an air stream (two-dimensional) at a velocity of v (or a circulation travels in air with a velocity of v). The *Kutta–Joukowski theorem* is: *A force/per unit length acting at right angles to the air stream, equals the product of the air density, the velocity of the air and the circulation.* The generated force is perpendicular to the airspeed and the direction of the circulation. The magnitude of the force F per unit length on the air can be determined by:

$$F = \rho v \Gamma \, [\mathrm{Nm}^{-1}] \tag{4.7}$$

where

Γ is the circulation in m^2s^{-1};
ρ is the air density in kgm^{-3}; and
v is the main stream air velocity in ms^{-1}.

Circulation Γ is an integral of the product of local velocity w and flow path dl along a closed loop Ω, and can be expressed by:

$$\Gamma = \oint_{\Omega} w\,dl \tag{4.8}$$

The direction of Γ follows the right-hand rule, shown in Figure 4.11. The direction of the force F in Equation (4.7) can be determine the direction of the force by another right-hand rule shown in Figure 4.12: stretch the thumb, index finger, and the middle finger of your right hand, and make

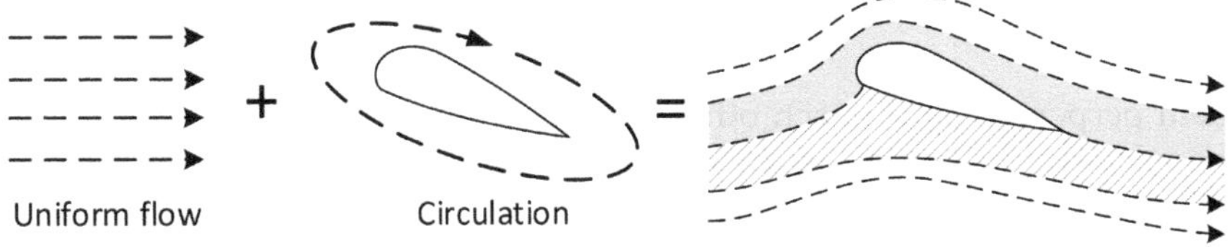

FIGURE 4.10
Illustration of circulation assumption of airflow around aerofoil.

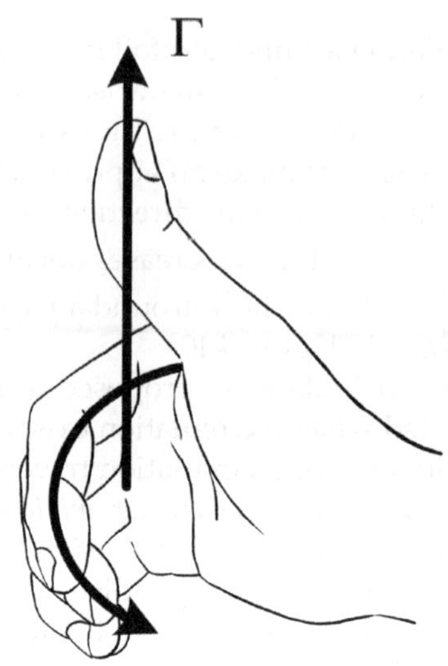

FIGURE 4.11
Direction of circulation.

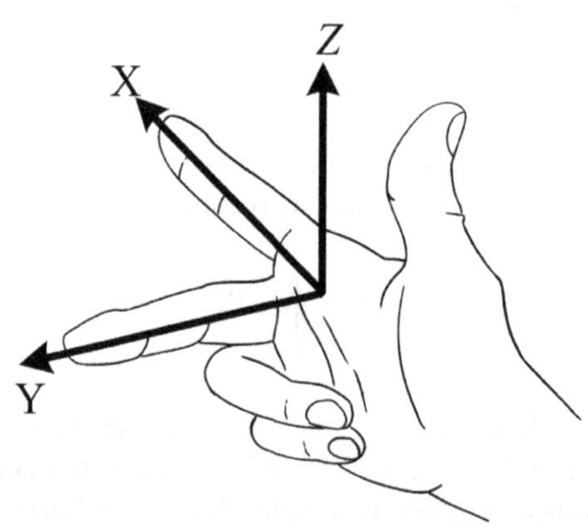

FIGURE 4.12
Direction of the force determined by the Kutta–Joukowski theorem.

them perpendicular to each other, in which "X" indicates the direction of the velocity v; "Y" indicates the direction of the rotation/circulation; and the "Z" indicates the direction of the force.

Applying the Kutta–Joukowski theorem to the airflow shown in Figure 4.10, the force acts on the aerofoil is vertically upward, that is *lift*.

This theory explained the upwash at the leading edge and downwash at the trailing edge by introducing fluid circulation, and quantified the lift produced by the aerofoil. If we follow a perimeter of the circulation shown in Figure 4.10 around the aerofoil with an approximation that the perimeter is approximately proportional to the circumference with the diameter of c, the chord of the aerofoil, the circulation around the aerofoil calculated by formula (4.8) would be:

$$\Gamma = \oint_{\Omega} w dl \propto \pi c w \tag{4.9}$$

Following the Kutta–Joukowski theorem, the lift on the aerofoil per span is:

$$L \propto \rho v w \pi c \ \left[\mathrm{Nm}^{-1}\right] \tag{4.10}$$

Comparing formula (4.1) to calculate the lift generated by an aerofoil (if the span of the aerofoil is b) and the lift obtained by the Circulation theory (4.10), the lift produced by both Equations (4.1) and (4.10) should be equal:

$$L = C_L \frac{1}{2} \rho v^2 S \propto \rho v w \pi c b = \rho v w \pi S \tag{4.11}$$

It is, then, obtained:

$$C_L \propto 2\pi \frac{w}{v} \tag{4.12}$$

It is noted that the ratio of $\frac{w}{v}$ changes with AoA of the aerofoil, α. It is shown in Figure 4.8 (b) and (d) that $\tan\alpha \approx \frac{w}{v}$. When α is relatively small, it can be assumed: $\tan\alpha \approx \alpha$, where α is in radians. So C_L in Equation (4.12) becomes:

$$C_L \propto 2\pi\alpha \ \ [\alpha \text{ is in radians}] \tag{4.13}$$

This expression (4.13) displays a linear relationship between the lift coefficient and the AoA. A typical curve of C_L vs. AoA for a moderate thickness cambered aerofoil (e.g. thickness:13% of c; camber: 2% of c) from reference AP3456 (AP3456, 1995) in Figure 4.13 shows this linear relationship. The lift coefficient increases proportionally with the angle of attack (AoA) within the valid range (AoA < 16°). Equation (4.13) was derived from the assumption of thin symmetrical aerofoil, which means that the lift coefficient is "0" when

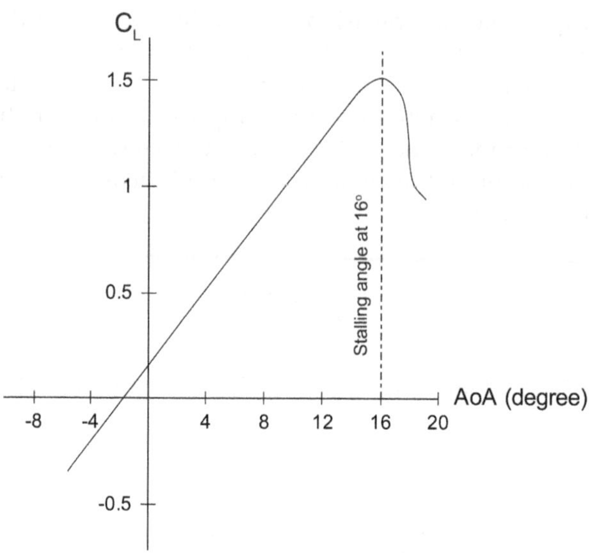

FIGURE 4.13
Change of lift coefficient with angle of attack (Credit: AP3456, MOD Crown copy right 1995, modified from 1-1-3 Fig 7 in AP3456).

the angle of attack is "0." It would not be true, if the real aerofoil is cambered. The (4.13) can be written if the aerofoil is cambered:

$$C_L = k_L(C_{L0} + 2\pi\alpha) \tag{4.14}$$

where C_{L0} is the coefficient of lift at AoA = 0. It is a function of the characteristics of the aerofoil camber. C_{L0} is "0", if it is a symmetrical aerofoil; k_L is a proportionality factor, a constant (empirical value), which is related to the features of aerofoil/wing, such as camber, thickness, aspect ratio.

Drag

Drag is another major component of the aerodynamic forces exerted on an aircraft when it is in flight. Drag is a force, so it is a vector. Its direction is the same as the relative air velocity of the aircraft. There are two types of the total drag the aircraft experiences in flight: *induced drag*, which is related to lift produced by the aerofoils, and *parasite drag*, which exists over the aircraft body due to the airflow around the aircraft. Understand the origin, or the cause of different parts of the drag can lead pilots controlling the aircraft more effectively.

Induced Drag

Airflow around an aircraft wing is three-dimensional (3-D), because all of real wings used in an aircraft are with finite spans as shown in Figure 4.4. The air pressure over an aerofoil is lower than the pressure in free-stream, in particular, at the wing root. The pressure difference will drive the air over the wing to flows inwards: the spanwise component of the air velocity heads toward the fuselage. On the other hand, the air pressure below the aerofoil is likely greater than the pressure in free-stream, and this pressure difference will make the air flow outward: the spanwise component of the air velocity heads away from the fuselage, as shown in Figure 4.14 (b). The two airflow streams above the wing and below the wing meet at the trailing edge of the wing.

As air is a viscous fluid, there is a tendency between the air particles to drive the air particles to reduce the difference in their movements to move together. So the air particles encounter and interact, causing rotating motion at the trailing edge, with the result being that *trailing edge vortices* are formed.

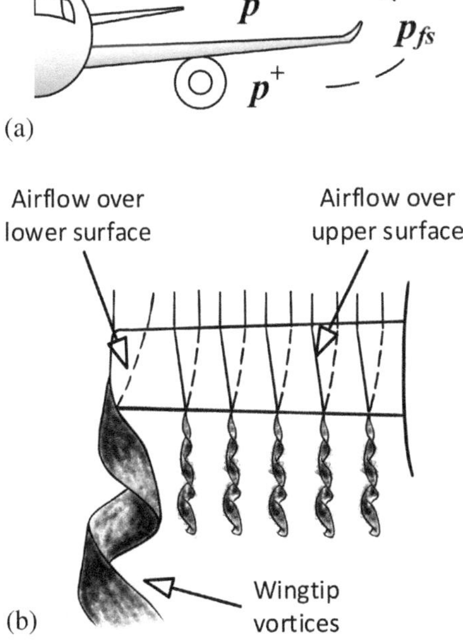

FIGURE 4.14
(a) Difference in pressure between over the upper surface and under lower surface of a wing; (b) Trailing edge vortices and wingtip vortex.

At the wingtip, apart from airflow going in different directions between upper and lower surfaces of the wing, there is a relative strong air pressure gradient. The air pressure at upper surface, p^-, of the wingtip is lower than the free-stream pressure and the pressure at the lower surface of the wingtip, p^+, is higher than the free-stream pressure. The differences in pressure at the wingtip drive the airflow at the wingtip into rotating motion – provide the wingtip airflow with a torque. As the result, a strong vortex is formed – *wingtip vortex*.

The trailing edge vortices and a wingtip vortex of a wing are shown in Figure 4.14 (a). Strong wing tip vortices can be seen behind an aircraft in air in Figure 4.15. The pressure at the center of a vortex decreases significantly, which can make the water vapor in the air condense, and we can see the "white trail" of the vortices.

The pressure difference between the air pressure in the center of a vortex and the pressure outside the vortex enhance the downwash, and then induce the airflow passing around the wing changing its direction to be close to the wing, as shown in Figure 4.16. There will be a stronger downwash, as

FIGURE 4.15
Visible wingtip vortices of an aircraft (Credit: NASA).

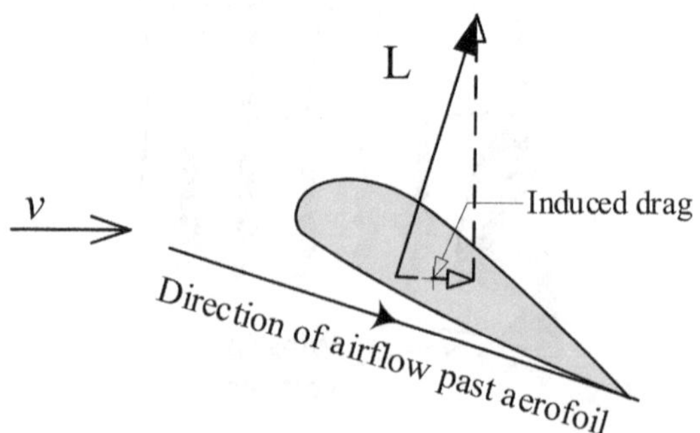

FIGURE 4.16
Illustration of the definition of induced drag.

the intensity of the vortices increase. The change in direction of the airflow reduces the effective angle of attack of the aerofoil, and the lift will then be reduced. The lift is perpendicular to the direction of the relative airflow, and the vector of the lift will tilts backward when the direction of the relative airflow changes, seen in Figure 4.16. The tilted lift vector is the sum of two components in vertical and horizontal directions. The component in vertical direction is the "true lift", which overcomes the weight of the aircraft. The magnitude of this "true lift" is less than that of the tilted lift. The component in the horizontal direction to the aircraft is a drag that is the *induced drag*. Therefore, the induced drag is, usually, said as a part of lift.

The amount of lift to balance the weight has been reduced due to the change of the airflow direction and the tilted vector of the lift produced by the aerofoil. To restore the amount of lift required, the angle of attack has to be increased. Furthermore, there will be more induced downwash, and lift vector will be induced to tilt back more, and as the result, the induced drag will be increased further.

Induced Drag Coefficient

Induced drag is a part of the tilted lift induced by downwash/vortices. In the Circulation theory of lift, the expression of lift of a symmetrical thin aerofoil can be derived from the Kutta–Joukowski theorem. In turn, the induced drag should be derived from the same theorem. The induced drag caused by the downwash, so the downwash (velocity) w and the circulation Γ, produce a force F_w given by the Kutta–Joukowski theorem:

$$F_w = \rho w \Gamma, \,[\mathrm{N/m}] \tag{4.15}$$

Use the right-hand rule shown in Figure 4.12 to determine the direction of F_w (Z), which is perpendicular to w the downwash velocity (X), and in the same direction as that of the relative air speed, or opposite to the movement of the aircraft, so it is a drag per wing span, i.e. the induced drag per wing span. Applying the approximation of the circulation about the aerofoil (4.10) in Equation (4.15), it becomes:

$$F_w \propto \pi c \rho w^2 \,[\mathrm{N/m}] \tag{4.16}$$

which demonstrates the nature of the induced drag – it is caused by the induced downwash. Combining Equations (4.2) and (4.16), the induced drag D_{in} over the wing is:

$$D_{\mathrm{in}} = C_{\mathrm{Din}} \frac{1}{2} \rho v^2 S = F_w \propto \pi c \rho w^2 b \,[\mathrm{N}] \tag{4.17}$$

where $S = bc$.

Reorganize Equation (4.17) and the expression of the coefficient of the induced drag can be,

$$C_{\text{Din}} \propto \frac{w^2}{v^2} \approx \alpha^2 \tag{4.18}$$

This assumed aerofoil is a symmetrical thin aerofoil, so to use Equation (4.13) rewrite Equation (4.18):

$$C_{\text{Din}} \propto C_L^2 \tag{4.19}$$

A real common wing has a wing tip, and very likely is cambered and not very thin. The airflow around it will be three-dimensional. So Equation (4.19) needs to be modified to consider the geometric features of aerofoil. The commonly used, theoretical based and empirical expression of the coefficient of induced drag, which can be found in training manuals (Robson, 2010; Boys and Wagtendonk, 2011) is:

$$C_{\text{Din}} = k\frac{C_L^2}{\pi AR} \tag{4.20}$$

or:

$$C_{\text{Din}} = \frac{C_L^2}{\kappa\pi AR} \tag{4.21}$$

where AR is the aspect ratio, and k, or κ are constants related to the type of aerofoil, and the shape of wing planform. For the design of aerofoil and wing planform, which could maintain close to 2-D airflow over the wing, k, and κ are "1"; for the design of aerofoil and wing planform, which produce strong 3-D airflow over the wing, $k > 1$, or $\kappa < 1$.

The coefficient of induced drag increases with the angle of attack, according to (4.20) or (4.21). For example, an aircraft requires extra amount of thrust at the taking-off phase of a flight, because taking-off need a greater angle of attack in order to have a high lift coefficient, which in turn, produces a high induced drag. The coefficient of induced drag will also be high if the airflow is diverged from two-dimensional flow. A high aspect ratio wing will make the airflow over the wing more close to two-dimensional airflow. Any feature of aerofoil, which enhances the three-dimensional airflow over the aerofoil, will produce a high value of coefficient of induced drag.

Example 4.2

The area of wing of an aircraft is 36 m^2, and its aspect ratio $AR = 6.25$. The shape constant, $k = 1$. Assume the air density = 1.225 kgm^{-3}. Calculate induced drag coefficient, induced drag and the power required to overcome the induced drag, when the aircraft is (a) at low speed, 90 kt, and the lift coefficient $C_L = 1.1$; and (b) at higher airspeed, 190kt, and its AoA = 2.6°.

SOLUTION

(a) The airspeed = 90 × 0.514 = 46.26 ms^{-1}. Use Equation (4.22) to calculate the coefficient of induced drag:

$$C_{\text{Din}} = k\frac{C_L^2}{\pi AR} = 1 \times \frac{1.1^2}{3.14 \times 6.25} = 0.062.$$

Use Equation (4.2) to calculate the induced drag:

$$D_{\text{in}} = C_{\text{Din}}\frac{1}{2}\rho v^2 S = 0.062 \times \frac{1}{2} \times 1.225 \times 46.26^2 \times 36 = 2925.6 \text{ N}$$

The power required to overcome the induced drag P_{in} is

$$P_{\text{in}} = D_{\text{in}}v = 2915.6 \times 46.26 = 135.3 \text{ kW}.$$

(b) The airspeed = 190 × 0.514 = 97.66 ms^{-1}. Use (4.16) to determine the lift coefficient C_L:

$$C_L = 2\pi\alpha = 2 \times 3.14 \times 2.6 \times 3.14/180 = 0.285.$$

Use Equation (4.22) to calculate the coefficient of induced drag:

$$C_{\text{Din}} = k\frac{C_L^2}{\pi AR} = 1 \times \frac{0.285^2}{3.14 \times 6.25} = 0.0041.$$

Use Equation (4.2) to calculate the induced drag:

$$D_{\text{in}} = C_{\text{Din}}\frac{1}{2}\rho v^2 S = 0.0041 \times \frac{1}{2} \times 1.225 \times 97.66^2 \times 36 = 869.4 \text{ N}$$

The power required to overcome the induced drag P_{in} is

$$P_{\text{in}} = D_{\text{in}}v = 869.4 \times 97.66 = 84.9 \text{ kW}.$$

Parasite Drag

Parasite drag exist on the surface of the aircraft. It is closely related to the shape and quality of the surface of the whole aircraft, and it is "carried" by the aircraft itself. There are three components of drag as parasite drag: skin drag, form drag, and interference drag. Skin drag and form drag are *profile drag*.

Skin Drag

Air is a viscous fluid, as discussed in Chapter 3. On the surface of an aircraft, the speed of air particle is zero due to the viscosity of the air, and there is a boundary layer over the surface. Within the boundary layer, the air particle velocity changes significantly, so it produces viscous friction on the surface of the aircraft.

The viscous friction on the surface of an aircraft is the skin drag. The skin drag coefficient can vary with the airflow regime within the boundary layer. The regime of the boundary layer depends on the quality of the surface, i.e. material, the smoothness and the geometry. The magnitude of the skin drag coefficient is also a function of the Reynolds number of the flow. For example, the skin drag coefficient can be determined approximately by formulae (3.8) and (3.10) for laminar and turbulent boundary layer of a flat surface, respectively.

An unsmooth surface can trigger early formation of turbulent boundary layer. It draws more kinetic energy from the main stream of the airflow to increase the air particle speed within the boundary layer when it is formed, so the coefficient of skin drag of an unsmooth surface is relatively high.

Airspeed can change the Reynolds number of the air flow over an aerofoil, so when airspeed increases, Reynolds number increases as well, and the coefficient of skin drag, C_{Dskin}, will decrease lightly according to formulae (3.8) and (4.10) as shown in Figure 3.10, as long as there is no change in flow regime. According to the formula (4.2), which is used to calculate drag, the skin drag would increase with v^2, if the coefficient of skin drag were constant. So the overall effect of the combination of formulae (3.8), (3.10), and (4.2) is that skin drag increases with the increase of airspeed.

When an aircraft is in a maneuver, its angle of attack will change and this change might cause the change in flow regime, which might change the coefficient of skin drag. However, in general, the angle of attack does not directly affect the coefficient of skin drag.

Form Drag

Form drag is also can be called separation drag, because it is caused by the flow separation from any streamline flow, including boundary layer separation, as discussed in Chapter 3. When flow separation occurs, vortices will be produced and detached from the surface of the object, forming a vortex wake. The vortices cause a local lower pressure zone, so for the object traveling in air, the front pressure of the object is higher than the back pressure, which caused by the vortices. This pressure differential is a drag to the traveling object. The vortex wake draws the kinetic energy from the main airflow, so a stronger flow separation results in a higher form drag.

The intensity and extent of the vortex wake depends on the Reynolds number of the fluid flow, and the shape and surface roughness of the object. When the Reynolds number is large, the vortex wake is strong. Figure 3.14 shows the flow separation behind a cylindrical object increases when the Reynolds number increases. The separation is strong if the shape of the object is blunt, like the cylindrical object and the vertical disk shown in Figure 3.15. When the shape of the object is more streamlined and slender, like the horizontal plate, and the thin aerofoil in Figure 3.15, the flow separation is weak. So the more streamline shaped object will have a lower coefficient of form drag.

Form drag is a component of profile drag. Usually it is difficult to calculate the coefficient of form drag by analytical formula or equations. In practice, it can be obtained by tests in laboratories under assigned conditions with a series of values of the Reynolds number. Engineers traditionally use empirical formulae and design handbooks to find the values of the coefficients of form drag they need.

The coefficient of form drag of aerofoil will be affected directly by the true air speed (TAS), because the Reynolds number changes with TAS, and the intensity of the flow separation directly related to the Reynolds number. The coefficient of form drag of aerofoil can change with its AoA indirectly as well. When an aircraft flies at a constant airspeed and increases its AoA (e.g. in banking), the point of boundary layer separation will move forward. The intensity of the separation becomes stronger, so the form drag increases.

Interference Drag

Air flows pass different parts of an aircraft usually with different flow pattern. The airflow leaves a wake behind each part of an aircraft body. The parts, or components of the body of the aircraft can be wings, any flaps, undercarriage, engine nacelles, or any external accessories carried by the aircraft. Many of the components of aircraft mentioned earlier join one with another, for example, fuselage and wings, undercarriage and fuselage, or engine nacelles and wings. The airflows and the wakes can interact with each other at the joint areas. The airflows cannot proceed by following the original flow patterns, and they will intrude into each other's flow paths, and airflows can be compressed, which can cause local pressure change – pressure disturbance, turbulence, and even flow separation. As discussed above, pressure change, turbulent flow, and separation all result in increase of drag. So the regimes of the interacting airflows can affect the magnitude of the interference drag.

The wakes behind a part of aircraft body can be vortices, like wing tip/trailing edge vortices. The wakes can also be in the form of Karman vortex street. These wakes periodically detach from the bodies and propagate in the air around the aircraft. Some of the wakes can propagate on to another part of the aircraft body, and this encounter between a wake and solid part of the aircraft body can produce pressure disturbance waves, producing noise, even oscillation. For example, the turbulent wake from landing gear striking on flaps produces noise and vibration during take-off, as shown in Figure 4.7. The noise and oscillation dissipate a significant amount of energy, and produce drag to the aircraft. Interference drag can be present in any joint areas throughout the entire aircraft. Ideally, design of the body of an aircraft should result in a smooth pressure distribution at a joint zone between two different parts, or components of the aircraft body to avoid the constructive interference of pressure waves. (CFD simulations play a role in these aspects of design.)

In the effort of reducing interference drag of an aircraft, aircraft fairing techniques are used to provide the joint zone with a smooth and gradual transition in shape from a part of the body to another. Figure 4.17 (a) shows a wing-body fairing on a Jetstream J32 aircraft, and Figure 4.17 (b) shows the flap track fairings on a Jetstream J31 aircraft ZK-JSH. The flap track fairings are in a canoe-shape, which streamline and protect the flap operation mechanisms. Optimizing the lateral or longitudinal placement of components of aircraft, for examples, nacelles, wings, or empennage, and retractable undercarriage are all good design practice to reduce interference drag.

The Reynolds number/airspeed of airflow, the geometry design of an aircraft, and the type of maneuverer the aircraft is in air affect the coefficient of interference drag, because Reynolds number/airspeed determines the regimes of airflow and intensity of flow wakes, and the geometry design of aircraft and aircraft maneuverers affect how the interference should occur. The coefficient of interference drag cannot be obtained by a formula from a

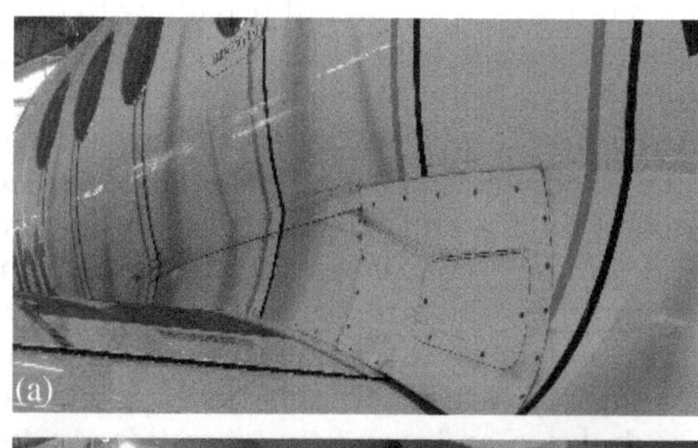

FIGURE 4.17
(a) Wing-body fairing (by author); (b) Flap track fairings (by author).

theory usually. It should be determined by tests of an aircraft in a wind tunnel, and by previous design data and experience, or good numerical (CFD) simulation, for example, the interference drag of a transonic wing with a high Reynolds number was estimated by numerical modeling and experiments reported in reference (Knight et al., 2011).

Total Drag

Total drag is the sum of induced drag and the various components of parasite drag. The coefficient of total drag is:

$$C_D = C_{\text{Din}} + C_{\text{Dskin}} + C_{\text{Dform}} + C_{\text{Dinterference}} \tag{4.22}$$

As each component varies with AoA, or airspeed, the total drag coefficient changes with angle of attack as well as changes with airspeed.

Figure 4.18 shows the changes of induced drag, parasite drag and the total drag of an aircraft with the airspeed. It is assumed that the aircraft is in level flight, which means $L = W$ (W is approximately constant within the concerned time period), so the airspeed increases and the lift coefficient needs to be decreased. The angle of attack will decrease for a lower lift coefficient, so $C_L \propto v^{-2}$. Then the coefficient of induced drag will be significantly decreased as $C_{\text{Din}} \propto C_L^2 \propto v^{-4}$. Therefore, the induced drag decreases with airspeed v, i.e. $D_{\text{in}} \propto v^{-2}$, as shown in Figure 4.18. On the other hand, the coefficient of parasite drag is related to the Reynolds number. The Reynolds number (*Re*) increases when airspeed increases, and the coefficient of the form drag and interference drag can increase with the Reynolds number, but the coefficient

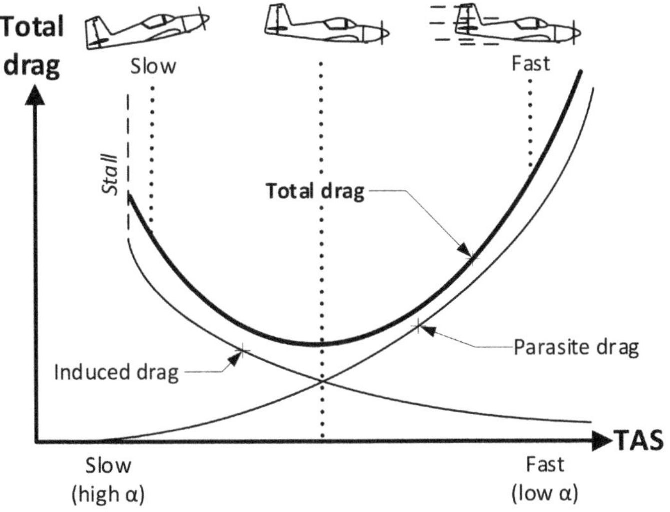

FIGURE 4.18
Drags vs. airspeed.

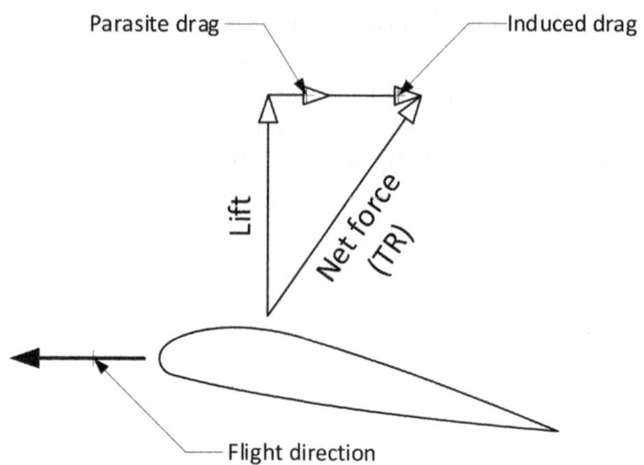

FIGURE 4.19
Vectors of lift and drags on aerofoil.

of skin drag can decrease with the increase of the Reynolds number. Overall, the parasite drag increases approximately with airspeed squared: v^2 in Figure 4.18. Figure 4.19 shows the vectors of aerodynamic forces generated on an aerofoil, where TR stands for total reaction.

Features of Aerofoil on Aerodynamic Forces

Aerodynamic force, lift, and drag are mainly produced by aerofoil. To produce high lift and low drag is the main target of aerofoil design. It is nearly impossible to design one aerofoil, which will produce highest lift coefficient and lowest drag coefficient with a high stall angle of attack. The physical features of an aerofoil, for example, the thickness, aspect ratio, camber, (and et al,) affect the aerofoil producing lift and drag. There are a wide variation of aerofoil designs. Some of them are for high payload; some of them are for high airspeed; some of them are for general usage; and others are for the aircraft with high maneuverer ability, for example, aerobatic performance.

In the previous sections of this chapter, we have analyzed the principles of producing lift, drag, and types of drag. In this section, we will apply this knowledge to understanding the effects of the physical features of aerofoil on aerodynamic forces.

Thickness of Aerofoil

We are interested here in the thickness of aerofoil, i.e. the relative thickness defined as the ratio of the thickness (t) to chord (t): t/c:

- The aerofoil is regarded as a thin aerofoil, if t/c is less than 7%. It suits high-speed aircraft.
- The aerofoil is regarded as general purpose aerofoil, if t/c is about 10% to 15%.
- The aerofoil is regarded as a thick aerofoil, if t/c is greater than 15%. It suits aircraft with a high payload.

For a thin aerofoil, the pressure differential between the upper surface and the lower surface of the aerofoil is relatively small, and its downwash is small as well. The vortices caused by the low pressure differential will be weaker, so the induced drag coefficient will be low. To compare with the aerofoil same "wing area" ($S = bc$), a thin aerofoil has less surface area, and will produce less skin drag. Therefore, a thin aerofoil design works well for a high-speed aircraft.

According to the Circulation theory of lift, the lower downwash of a thin aerofoil indicates a lower circulation. It means that the lift coefficient of this aerofoil will be relatively lower. Schematic diagram Figure 4.20 shows the difference in lift coefficient of two aerofoils with same camber, but different thickness, $(t/c)_1 > (t/c)_2$. It is clear that the aerofoil with a low relative thickness produces lower lift coefficient and has a lower stall AoA.

For a thick aerofoil, the downwash and the pressure differential are both greater than a thin aerofoil. So it will produce higher induced drag coefficient and higher lift coefficient, in particular, when AoA is relatively high. Stall is caused by an early boundary layer separation. One of the separation conditions is adverse pressure gradient. The condition of adverse pressure gradient occurs later for the thicker aerofoil if a thin and a thick aerofoil

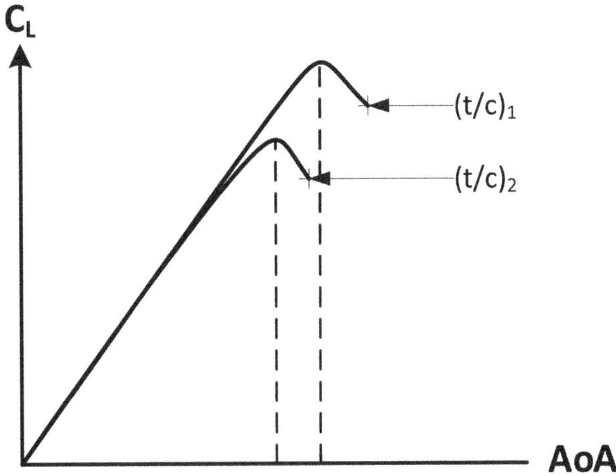

FIGURE 4.20
Lift coefficient change with thickness of aerofoil.

set in the same AoA, because the stronger induced downwash changed the direction of the relative airflow and decreases the effective AoA. This will lead to a higher stall AoA for the thick aerofoil. Normally the aircraft, which needs to produce high lift (high payload), will have a thick wing design.

Aspect Ratio of Aerofoil

Aspect ratio (AR) of an aerofoil is the ratio of wing span b to the chord c. The aspect ratio is a feature of the basic geometric shape of an airplane's planform. A high aspect ratio means the airplane with a long or narrow wing. For example, the aspect ratio of a glider can be as high as 33, and the aspect ratio of a passenger jet is about 10. A low aspect ratio means a short wing, or wide wing. For example, an aspect ratio of small subsonic aircraft can be approximately 5 or 6, and the aspect ratio of a jet fighter could be as low as 2 to 3.5.

The airflow around a wing with a low aspect ratio is three-dimensional. There is a strong spanwise component of air velocity over the wing, as shown in Figure 4.21 (a). The spanwise component of air velocity causes vortices behind the wing, so it will produce a higher induced drag. On the other hand, for a wing with a high aspect ratio, the airflow over the wing is more close to two-dimensional, less spanwise component, as shown in Figure 4.21 (b). It will produce weaker vortices and less induced drag.

The aspect ratio can affect the lift coefficient as well. The lift coefficient of a wing with a high aspect ratio is higher than that of a wing with a low aspect ratio, if they have the same relative thickness and same camber. The wing with a low aspect ratio should produce stronger downwash, which draws the relative airspeed closer to the wing, and induce the lift vector tilting rearward more, then the actual magnitude of lift is reduced. The effective angle

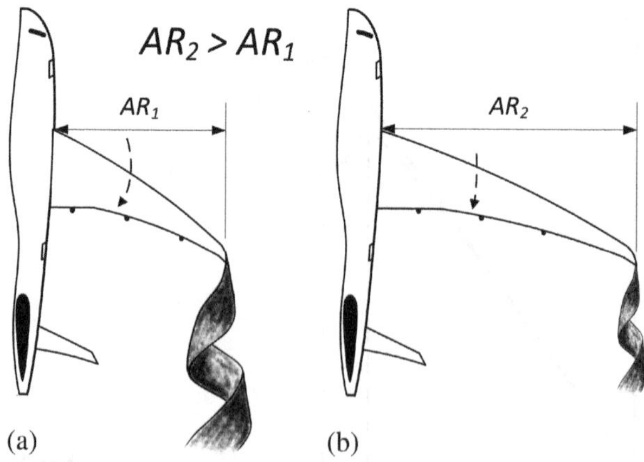

FIGURE 4.21
Airflow over aerofoil with different aspect ratios: (a) with a smaller aspect ratio; (b) with a greater aspect ratio.

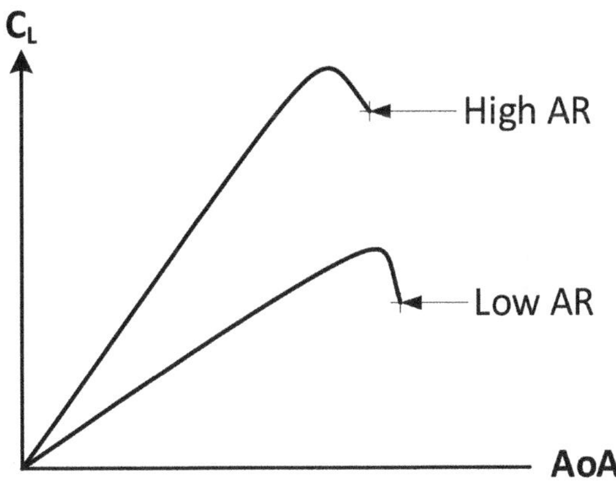

FIGURE 4.22
Lift coefficient change with aspect ratio.

of attack of this wing will be decreased as well. Therefore, the lift coefficient of the wing with a low aspect ratio should produce lower lift coefficient than the wing with a high aspect ratio, but the stall AoA of the wing with a low *AR* is higher than that of the wing with a high *AR* as shown in Figure 4.22.

Camber of Aerofoil

The common aerofoil used by the majority of airplanes is cambered. Cambered aerofoil can produce a higher lift coefficient C_L at the same AoA than a symmetric aerofoil does. As we know that $C_L > 0$ for a cambered aerofoil, while $C_L = 0$ for a symmetric aerofoil.

Considering two wings with same thickness and aspect ratio, one of which is cambered, and the other is symmetric, the circulation around the cambered wing would be higher at the same AoA than the other. A high circulation would lead to a higher lift coefficient. As shown in Figure 4.23, the lift coefficient increases with the increase of camber for the same AoA. If the lift coefficient is higher, the air airspeed required to produce same amount of lift at the same AoA could be lower from Equation (4.1). So the stall speed of an airplane with a cambered aerofoil would be lower than that with symmetric aerofoil.

On the other hand, an increase in camber will decrease the stall AoA of aerofoil. The adverse pressure gradient gets stronger after the maximum camber point over the aerofoil when the camber of the aerofoil increases. The strong adverse pressure gradient encourages the earlier boundary layer separation, which lead an early stall. So Figure 4.23 shows that the higher level of camber is, the smaller the stall AoA is.

The location of the maximum camber of aerofoil will affect the lift coefficient and the induced drag coefficient. The pressure distribution will change if the location of the maximum camber varies along its chord.

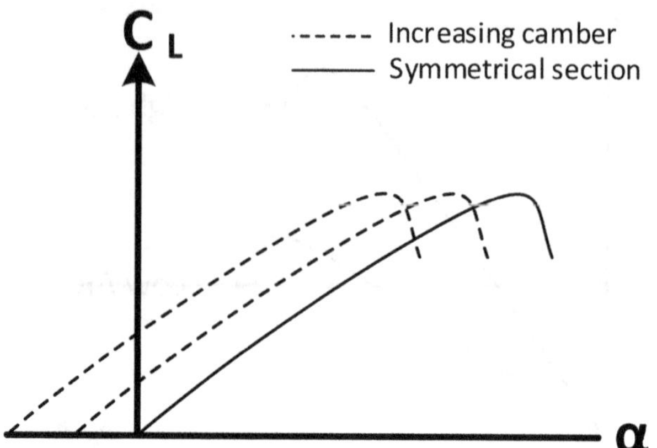

FIGURE 4.23
Lift coefficient change with different camber.

Laminar Flow Aerofoil

The design of laminar flow aerofoil (laminar flow in boundary layer) started before 1930, and more development of laminar aerofoil design was carried out in general aviation, e.g. work reported by Selig and CO. (Selig, Maughmer and Somer, 1995). As we discussed in Chapter 3, there are two regimes of fluid flow, laminar and turbulent. Laminar flow is a layered smooth fluid flow, and the viscous friction drag produced in a laminar boundary layer is lower than that in a turbulent boundary layer. So laminar flow over aerofoil should be an energy saving choice for an airplane.

The typical features of laminar aerofoil are thin leading edge, and further-back maximum camber, as shown in Figure 4.24. The thin leading edge can keep laminar airflow from the leading edge as long as possible, and it also ensures a gentle even change in air pressure over the aerofoil. The further-back maximum camber makes the adverse pressure gradient region over the aerofoil locate at the rear part of the aerofoil, which will delay the boundary layer separation. Therefore, the laminar flow design of aerofoil should have the advantage of low skin drag, low form drag, and increasing stall AoA.

Features to Delay/Prevent Boundary Layer Separation

Boundary layer separation is the cause of aircraft stall. As we have learned that boundary layer will separate if the air particles are stagnant, due to

FIGURE 4.24
Laminar flow aerofoil.

exhausting kinetic energy, and the local adverse pressure gradient. Therefore, boundary layer separation or stall will be delayed or prevented, if stagnant air particles, or local adverse pressure gradient, or both, can be eliminated or avoided.

There are many features on a wing with the function of delaying separation.

Suction: control the stagnant air particles in the boundary layer by suction. It can use a vacuum pump to achieve this, but the pump consumes power and add weight to the aircraft.

Blowing away, as shown in Figure 4.25 (a) and (b): to control the stagnant air particles in the boundary layer by blowing/pushing. A jet of air can be released from a vent over the wing to push the stagnant air particles toward trailing edge. It can be a part of bleed air or MEMS (See Figure 4.25 (a)) – micro-electro mechanical system. This system is an air cell with an orifice open on wing surface. There is a Piezo sensor attached to a diaphragm of this

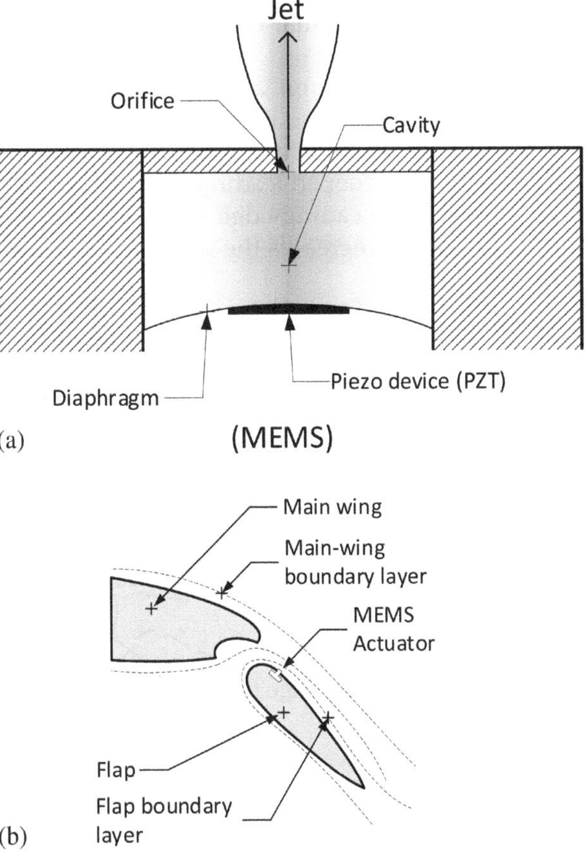

FIGURE 4.25
Boundary layer control: (a) MEMS – micro-electromechanical system; (b) blow-off.

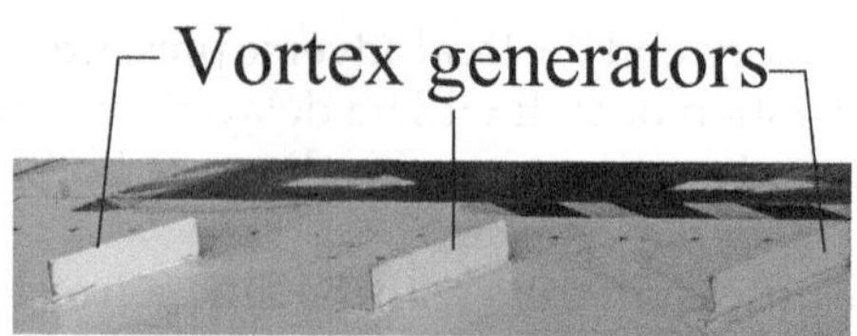

FIGURE 4.26
Vortex generators on an airplane.

cell. When this sensor receives an electronic signal, the diaphragm moves up to press the air out through the orifice.

Vortex generators, as shown in Figure 4.26: To produce micro-vortices to transfer more kinetic energy from the main stream of airflow to the boundary layer. When the boundary layer obtains more kinetic energy, it does not separate easily. It will increase skin drag, but it can increase the stall AoA, and make the wing not stall easily. There are different vortex generators with different shapes and sizes, and similar devices located at different parts of wings for example, vortilons.

There are many other devices on a wing, whose functions include increasing lift, and delaying stall, or increasing drag when it is required in flight, for example, the leading-edge slat, and trailing edge flap of a wing. Figure 4.27 shows the drooped leading-edge slat and extended trailing edge flap on a wing. The leading-edge slat extends when the aircraft takes off. This extension will increases the wing area to provide more lift,

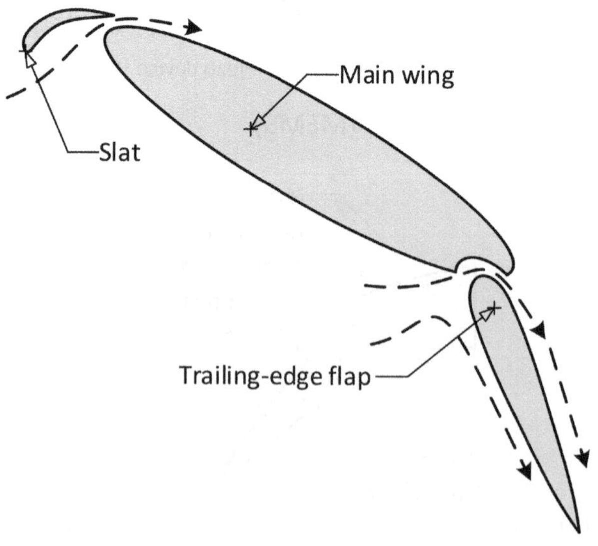

FIGURE 4.27
Schematic diagram of leading-edge slat and extended trailing edge flap.

and it also form a leading-edge slat, as shown in Figure 4.27. This slot allows a stream of airflow pass from the lower part of the leading edge through the slot to the upper surface of the wing. This air stream will increase the kinetic energy in the boundary layer over the wing, so it can delay the boundary layer separation/stall, while the aircraft is in a high AoA during taking-off phase of flight. The extended trailing edge flap will increase the effective AoA, i.e. increase the lift coefficient of the wing, as well as increase the wing area so the wing will increase lift significantly at taking-off.

Shape of Wing Tip

Induce drag, which is also called *vortex drag*, is caused by vortices at the trailing edge and wing tip. The wing tip vortex is stronger than that of trailing edge vortices. To reduce the intensity of wing tip vortices will proportionally reduce the induced drag. The function of wing tip devices is to reduce the intensity of wing tip vortex. Figure 4.28 shows some of the wing-tip devices.

A winglet is a commonly used wing tip device. A winglet can reduce the intensity of wing tip vortex by changing the pressure distribution around wingtip, and reforming the vortex into smaller vortices behind the winglet. The structure of a winglet is relatively simple, and it works effectively. There are varieties of wingtip devices, for example, up-winglet, down-winglet, wing fence, wingtip tank, and so on.

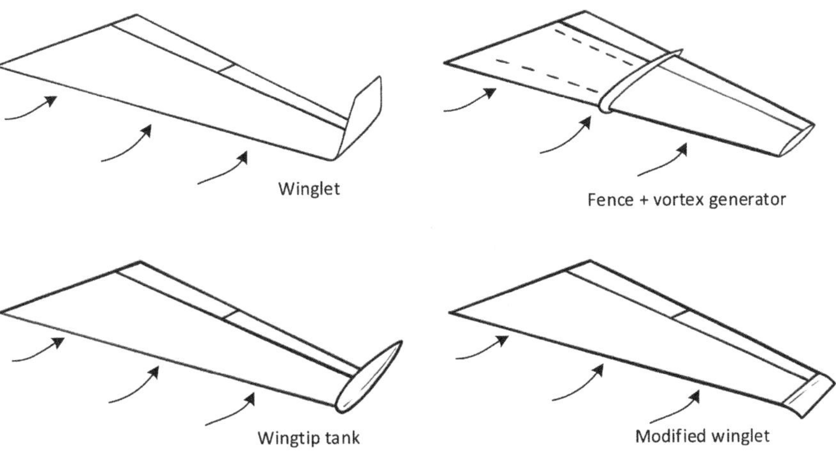

FIGURE 4.28
Wingtip devices.

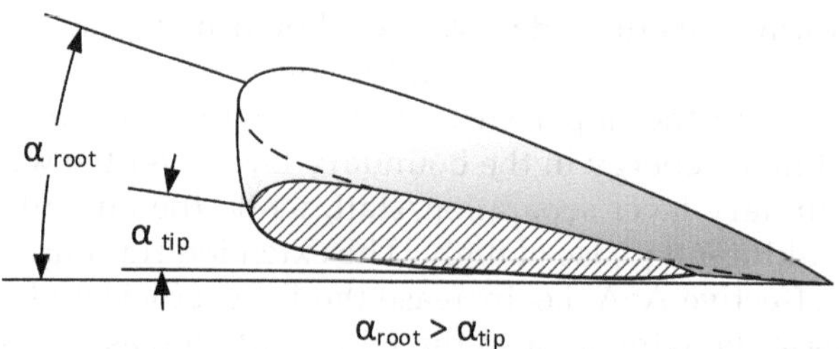

FIGURE 4.29
Washout feature.

Washout

The design of a wing with a feature that the incident angle at the wing root is greater than the incident angle at wingtip is called *washout*, shown in Figure 4.29. The incident angle gradually decreases from the wing root to wingtip. This feature will lead to a decrease in the pressure difference between the lower and upper surfaces at the wingtip to reduce the intensity of wingtip vortex, which, in turn, reduces the induced drag.

The reduced wingtip incident angle can delayed the adverse pressure gradient occurring, so it delays the wingtip stall, which should prevent the incident of wing drop, and ensure that the aileron on the wing can operate effectively.

Exercises

1. Aerofoils NACA3306 and NACA2412: which of them should have relatively higher lift coefficient?
2. What type of airflow is over a finite wing?
3. Why is the coefficient lift of a cambered aerofoil "0"?
4. What is induce drag?
5. Why is it said that the induced drag is part of lift?
6. State the effects of *AR* on coefficients of lift and drag.

5

Stability

When an aircraft is in a disturbance in flight, the aircraft will be in unexpected rotating motions. The aircraft is stable if it is able to produce a correct amount of moment in a correct direction to overcome the disturbance to restore the equilibrium. To discuss the stability of an aircraft is to discuss the ability of the aircraft to generate the restoring moment, which is produced by aerodynamic forces. So, first we should remember the formulae to calculate aerodynamic forces as well as some principles related to moment.

Revision on Moment

Aerodynamic Forces

Lift: produced by aerofoil, and can be calculated by Equation (4.1):

$$\text{Lift} = C_L \cdot \frac{1}{2}\rho v^2 \cdot S$$

Drag: aerodynamic resistance force produced in flight, can be calculated by Equation (4.2):

$$\text{Drag} = C_D \cdot \frac{1}{2}\rho v^2 \cdot S$$

where C_L, and C_D, are lift and drag coefficients, respectively; ρ is air density; S is the area of aerofoil; and v is airspeed.

Principles of Moment

Aircraft can rotate about three axes: pitching – rotating about lateral axis; rolling – rotating about longitudinal axis; and yawing – rotating about directional axis, as shown in Figure 5.1. To calculate the moments, which cause the rotating motions, following the formulae below:

Pitch Moment:

$$\text{Pitching Moment} = C_m \cdot \frac{1}{2}\rho v^2 \cdot S \cdot c \tag{5.1}$$

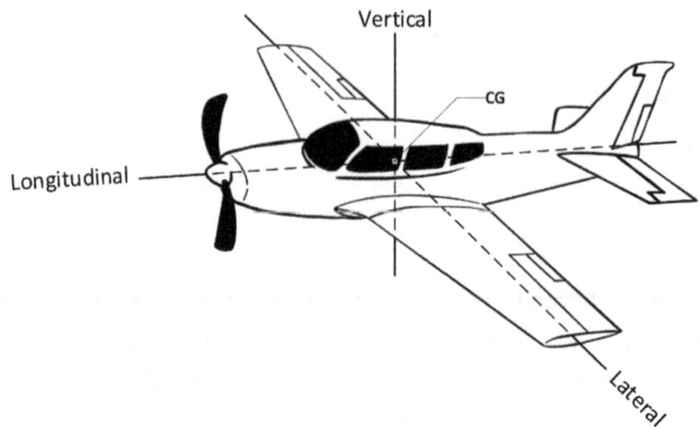

FIGURE 5.1
Rotational motion of aircraft in three directions.

where C_m is the coefficient of pitch moment, and c is the chord of the aerofoil. Pitch moment results in rotating motion about lateral axis, and the rotating motion is controlled mainly by the elevator.

Roll Moment:

$$\text{Rolling Moment} = C_R \cdot \frac{1}{2}\rho v^2 \cdot S \cdot b \tag{5.2}$$

where C_R is the coefficient of rolling moment, and b is the wingspan. Rolling moment results in rotating motion about longitudinal axis, and the rotating motion is controlled mainly by ailerons.

Yaw Moment:

$$\text{Yawing Moment} = C_Y \cdot \frac{1}{2}\rho v^2 \cdot S \cdot b \tag{5.3}$$

where C_Y is the coefficient of yawing moment. Yawing moment results in rotating motion about vertical axis, and the rotating motion is controlled mainly by the rudder.

Moment

The magnitude and direction of a moment m produced by a force depend on the magnitude of the force and the relative position of the force to the pivot point, or the rotating center. As shown in Figure 5.2 (a), a force **F** acts on one end of a stick vertically; the length of the stick is l, and the pivot is "O" at the other end of the stick. d is the (perpendicular) distance between the force

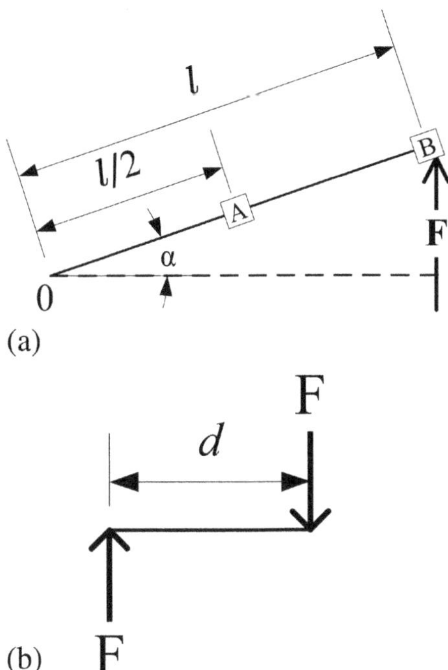

FIGURE 5.2
(a) A single force system; (b) A force couple system.

and the pivot point. The angle between l and d is α. The moment caused by F about "O" is:

$$M = Fd = Fl\cos\alpha \tag{5.4}$$

where F is magnitude (N) of the force **F**, and the unit of moment M is Nm. The moment caused by **F** will change, if the chosen pivot point is different from "O".

Example 5.1

Find moment by the force **F** about pivot O, A, and B, respectively, in the force system shown in Figure 5.2 (a). l is 2 m, and H is located at the half-way point between O and B, while the force **F** acts on B. The magnitude of the force is 20 N, and the force acts vertically upward. The angle α is 20° from horizontal level.

SOLUTION

Moment about pivot O: Use Equation (5.4):

$M_O = Fl\cos\alpha = 20 \times 2 \times \cos 20° = 37.6$ (Nm); "+" anticlockwise

Moment about pivot A: the distance from A to B, where the force acts on, is $l/2$, Use Equation (5.4):

$$M_A = F\frac{l}{2}\cos\alpha = 20\times1\times\cos 20° = 18.8\ (\mathrm{Nm});\ \text{“+” anticlockwise}$$

Moment about pivot B: The force is on the pivot point, so there is no distance between the pivot point and the force:

$M_B = Fd = 20\times0 = 0$. The moment is "0".

Force Couple

A force system that consists of two forces of equal magnitude F in opposite directions, shown in Figure 5.2 (b), is called *force couple*. If the distance between these two forces is d, the moment of this system is $M=Fd$, which is independent of the location of pivot point. So the total force of a force couple system is "0", but the moment of this system is not "0". So the force couple system is not in equilibrium.

Superposition of Force

A force in a system can be moved from one location to another without changing the total force and the total moment of the system by using the principle of superposition. In Figure 5.3, the force system at the left shows that a force **F** acts upward at point A, and the distance from **F** to point B is d. The total force of this system is F and the total moment about pivot point B of this system is Fd.

A pair of forces with the same magnitude as F can be added at point B without changing the nature of the original system, as shown in Figure 5.3. However, **F** at A and the negative force at B form a force couple, and the positive force at B has the same magnitude and the direction as **F** does. The new equivalent system consists of a force couple and an upright force at point B. The total force of the new system is **F**, and the total moment about pivot point B of this new system is Fd – moment of the force couple. So the new system is the same as the original system – *superposition.*

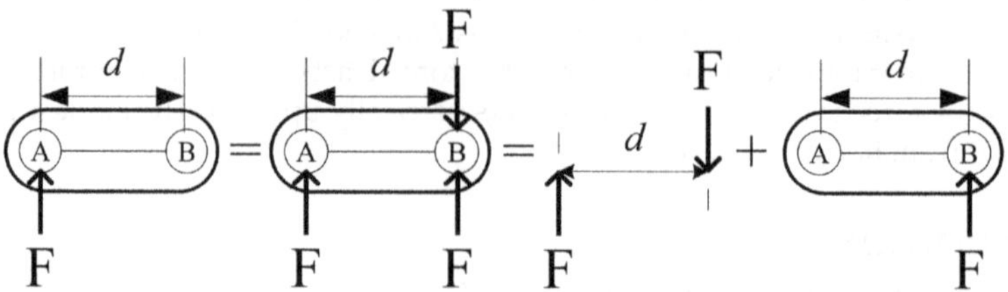

FIGURE 5.3
Force system superposition.

Example 5.2

Find the moment about point A in the original force system at the left on Figure 5.3 and in the system after the superposition at the right side in Figure 5.3. A and B are on the same horizontal line, and the distance between A and B is 0.5 m. The force **F** acts on A vertically. The magnitude of the force is 20 N.

SOLUTION

Moment about A in the system on the left: the force is on the pivot point, so there is no distance between the pivot point and the force:

$$M_A = Fd = 20 \times 0 = 0;$$

Moment about pivot A in the superposition system: the moment by the force couple + the moment about A by F at B:

$$M_A = -Fl\,(\text{clockwise}) + Fl = -20 \times 0.5 + 20 \times 0.5 = 0;$$

The moment about A in both systems is the same.

Therefore, in the practical use of superposition principle, any force in a system can be transferred artificially from its acting point A to a new location B with the inclusion of a force couple. The nature of the new force system is the exact same as the original force system. The moment of this force couple is equal to the moment by the force in the original system with the new location B as the pivot point.

Pitch Moment and Pitch Moment Coefficient

When an aerofoil sits with an angle of attack (AoA) to the relative airflow, there is a pressure distribution around the aerofoil, and this pressure distribution changes with AoA, as shown in Figure 5.4. The pressure distribution is in fact the distribution of the relative pressure: $(p - p_{fs})$, the difference between the local pressure and the free-stream pressure. The profile of the relative pressure distribution changes with AoA of the aerofoil. The solid line indicates that the local relative pressure is positive, while the dashed line indicates that the local relative pressure is negative in Figure 5.4. Integrating the pressure difference over the whole surface of aerofoil, a total lift can be obtained as shown by the vertical solid arrow in Figure 5.4, which acts at the center of pressure (CP) of the aerofoil. The sum of lift and drag on the aerofoil is the total aerodynamic force, called *Total Reaction* (TR) of the aerofoil. TR is on CP, whose location changes with AoA along the chord of aerofoil.

We can use the Equation (5.1) to obtain the pitch moment produced by aerodynamic forces. The pitch moment is positive "+" if the pitch moment would drive the aircraft nose-up, and the pitch moment is negative "–" if the pitch moment would drive the aircraft nose-down. As we know, the pitch

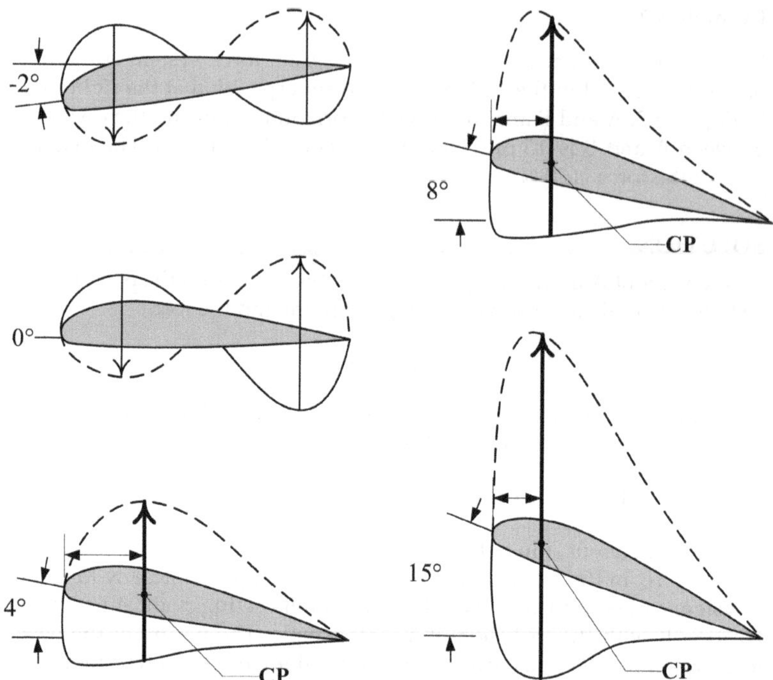

FIGURE 5.4
Change of pressure distribution with angle of attack (AoA).

moment will be different if the pivot point of pitching is at a different location. The coefficient of pitch moment, C_m, in Equation (5.1) represents the difference. For example, the coefficient of pitch moment about leading edge of aerofoil, C_{mLE} is negative when AoA is positive, while the coefficient of pitch moment about leading edge of aerofoil, C_{mLE} is positive, when AoA is positive, as shown in Figure 5.5, which is based on the data of RAF 15, in Kermode (2012). The pitch moment about CP (center of pressure) is "0", i.e. $C_{mCP}=0$.

The coefficient of pitch moment C_m changes with AoA due to the change of pressure distribution and CP. According to Equation (5.1), pitch moment changes with airspeed as well as with AoA.

The total lift produced is "0" when the angle of attack is −2° for a cambered aerofoil shown in Figure 5.4. This angle of attack is called *zero-lift angle*. The zero-lift angle for a cambered aerofoil is not 0°. At this angle of attack, the magnitudes of the positive lift and the negative lift are the same, but they situate at different locations on the chord, so they act as a force couple, and there is a moment at zero-lift angle. As we know, the moment of a force couple is independent of location of pivot point, so the moment coefficient about LE, TE, and AC (aerodynamic center) will be the same at the zero-lift angle. For a symmetrical aerofoil, the zero-lift angle is at AoA = 0°, and there is no moment, because the relative pressure on upper and lower surface at the same location on the chord is the same.

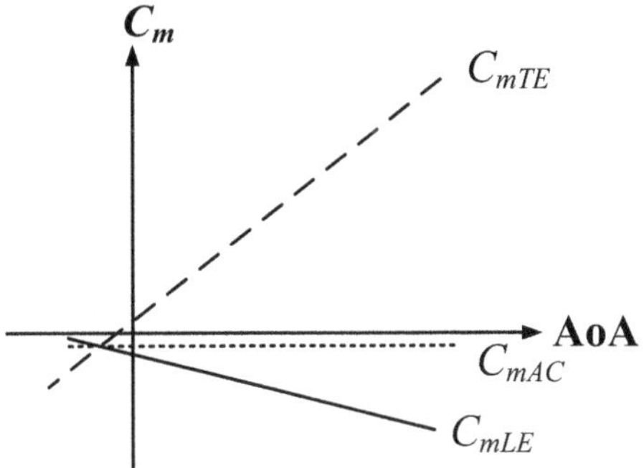

FIGURE 5.5
Schematic diagram of pitch moment coefficient about LE, TE, and AC with AoA.

Aerodynamic Center (AC)

There is a special point along a chord of an aerofoil about which the pitching moment approximately does not change with angle of attack, which means that C_{mAC} does not change with AoA; and the moment remains as the relatively small negative value at the "0" lift angle if the airspeed is constant. This point is called *aerodynamic center.* Typically, the aerodynamic center lies at about a quarter of the chord length from the leading edge (LE), and it is a feature of the aerofoil.

At the zero-lift angle, the total lift is "0" but the pitch moment is not. If the drag is relatively small, the moment at the zero-lift angle would be dominated by the moment produced by the lift force couple, which is independent of the location of the pivot point. So the pitch moment at the zero-lift angle could be approximately as the same as the pitch moment about the AC of the aerofoil, which is nearly independent of AoA. We can move the aerodynamic forces to the AC, when we analyze the effect of AoA change on pitch moment easily, since the location of AC is a feature of aerofoil, and the moment about the AC does not change with AoA.

Calculation of Aerodynamic Center

Where the aerodynamic center exactly lies on an aerofoil is dependent on the design of the aerofoil. There are a number of ways to calculate the location of AC on the chord. In this chapter, the method to find the AC is to use superposition concept with the information of lift and drag coefficients of the aerofoil. The method is explained in example 5.3.

Example 5.3

The AoA, α, of the aerofoil shown (not to scale) in Figure 5.6 to the relative airflow is 8° and chord of the aerofoil is 2 m. TR acts at the center of pressure on the chord, expressed as the sum of lift L and drag D. For this aerofoil (RAF-15) at $\alpha = 8°$, $C_L = 0.76$, $C_D = 0.044$, $C_{m,LE} = -0.24$, $C_{m,TE} = 0.53$, and $C_{m,AC}=-0.061$. Find the location of the aerodynamic center (AC) of this aerofoil, and the center of pressure (CP) at this AoA along the chord. (Data from Kermode, 2012)

SOLUTION

To find out the location of AC, use the method of calculating pitch moment about AC. TR is at the CP, but the location of CP on the chord is unknown. TR should be transferred to a known location, for example, to the leading edge (LE) or the trailing edge (TE).

Firstly, assume that the distance from the LE to the aerodynamic center (AC) is x, and move TR, or L and drag D to LE. This move of L and D generates a moment: the pitch moment about LE, M_{LE}, according to the superposition concept, shown in Figure 5.7. The moment about aerodynamic center when TR is at CP originally and the pitch moment about AC when TR is moved to LE should be the same. The pitch moment about AC in the original force system is M_{AC}, and the pitch moment about AC in the moved force system consists of M_{LE}, the moment by L about AC, M_L, and the moment by D about AC, M_D. Then we have the following equation of the pitch moment about AC:

$$M_{AC} = M_{LE} + M_L + M_D \tag{5.5}$$

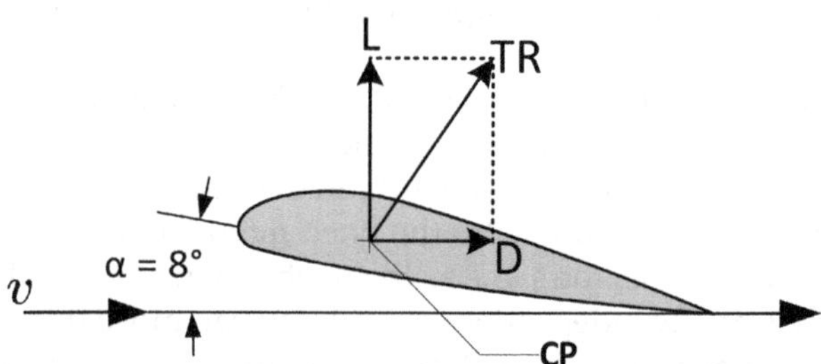

FIGURE 5.6
Lift and drag on CP when AoA is 8°.

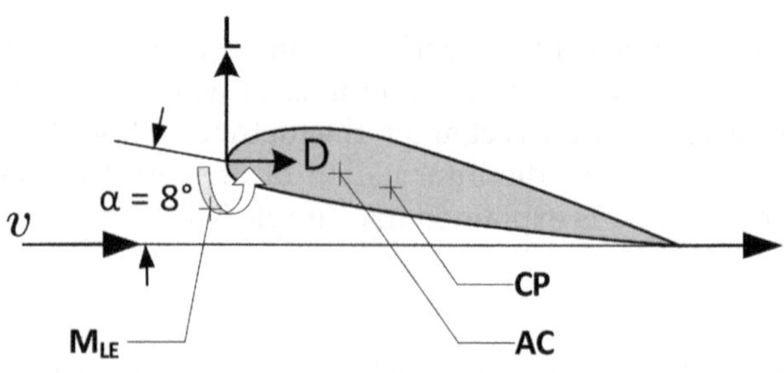

FIGURE 5.7
Lift and drag are relocated to leading edge, LE.

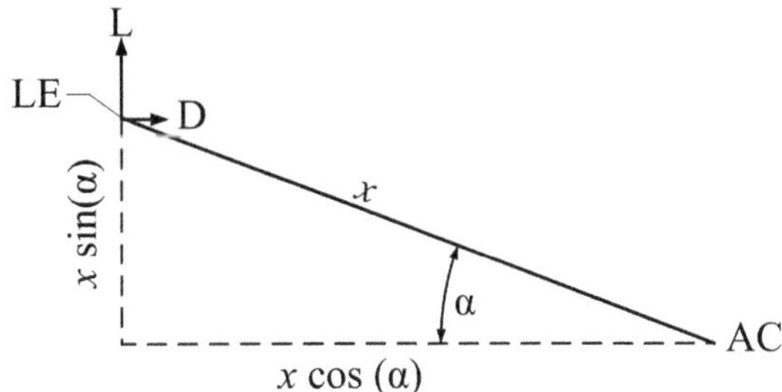

FIGURE 5.8
Distance triangle with L and D at LE, to aerodynamic center, AC.

The distance from L to AC is $x \cdot \cos 8°$, and the distance from D to AC is $x \cdot \sin 8°$ shown in Figure 5.8. Using Equation (5.1) and (5.4):

$$M_{AC} = C_{m,AC}\frac{1}{2}\rho v^2 Sc$$

$$M_{LE} = C_{m,LE}\frac{1}{2}\rho v^2 Sc$$

$$M_L = C_L\frac{1}{2}\rho v^2 Sx\cos\alpha$$

$$M_D = C_D\frac{1}{2}\rho v^2 Sx\sin\alpha$$

So Equation (5.5) is:

$$C_{m,AC}\frac{1}{2}\rho v^2 Sc = C_{m,LE}\frac{1}{2}\rho v^2 Sc + C_L\frac{1}{2}\rho v^2 Sx\cos 8° + C_D\frac{1}{2}\rho v^2 Sx\sin 8° \quad (5.6)$$

Simplify (5.6):

$$C_{m,AC}c = C_{m,LE}c + C_L x\cos 8° + C_D x\sin 8° \quad (5.7)$$

then:

$$-0.061c = -0.24c + (0.76 \times \cos 8° + 0.044 \times \sin 8°)x$$

$$\frac{x}{c} = \frac{-0.061 + 0.24}{0.76 \times \cos 8° + 0.044 \times \sin 8°} \doteq 0.236$$

$$x = 23.6\% \cdot c = 0.236 \times 2 \doteq 0.47\ (\text{m})$$

The aerodynamic center is 23.6% of c, the chord, i.e. 0.47 m from LE.

Secondly, TR can be moved to TE, and assume that the distance from AC to TE is x_T, as shown in Figure 5.9. Then, the process of calculation will be very similar to above. Equation (5.5) will become:

$$M_{AC} = M_{LE} - M_L - M_D \quad (5.8)$$

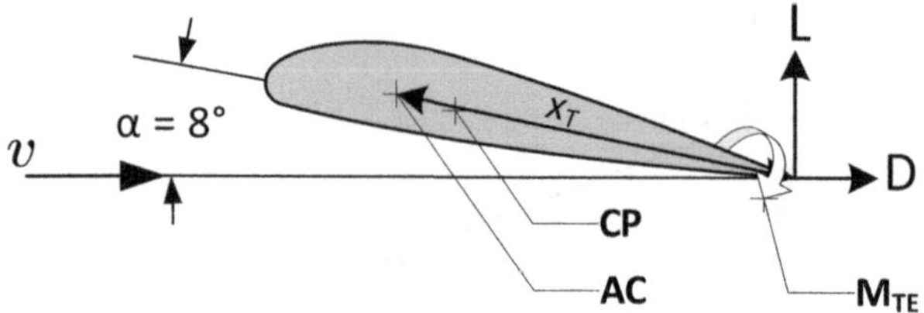

FIGURE 5.9
Lift and drag are relocated to trailing edge, TE.

Note that the pitch moments about AC by L and D at TE are nose-down moments. The Equation (5.6) becomes:

$$C_{m,AC}\frac{1}{2}\rho v^2 Sc = C_{m,TE}\frac{1}{2}\rho v^2 Sc - C_L\frac{1}{2}\rho v^2 Sx_T\cos 8^\circ - C_D\frac{1}{2}\rho v^2 Sx_T\sin 8^\circ \quad (5.9)$$

The solution of Equation (5.9) is:

$$x_T \doteq 76.4\% \cdot c = 0.764\times 2 \doteq 1.53\ (\text{m}).$$

The aerodynamic center is 76.4% of c, the chord, i.e. 1.53 m from TE, which agrees with the result of using LE. A similar result will be obtained if it is calculated with a different AoA.

The center of pressure (CP) at this AoA can be calculated by using the same method: moving TR to LE, and assume that the distance from LE to CP is y. The moment by TR about the CP should always be "0", so Equation (5.5) becomes:

$$0 = M_{LE} + M_L + M_D \quad (5.10)$$

Then Equation (5.7) becomes:

$$0 = C_{m,LE}c + C_L y\cos 8^\circ + C_D y\sin 8^\circ \quad (5.11)$$

$$\frac{y}{c} = \frac{0.24}{0.76\times\cos 8^\circ + 0.044\times\sin 8^\circ} \doteq 0.316$$

$$y = 31.6\% \cdot c \doteq 0.63\ (\text{m})$$

The center of pressure (CP) is 31.6% of c, the chord, i.e. 0.63 m from LE at AoA = 8°.

The value of y will change if the angle of attack is different. This feature of CP can be seen in Figure 5.4. However, AC and CP are always at the same location if the aerofoil is symmetrical. So for a symmetrical aerofoil, $C_{m,AC} = 0$.

Longitudinal Stability

When an aircraft is in equilibrium, the total force on the aircraft is "0" and the total moment in pitching rolling and yawing is "0" as well. It means that the total pitch moment about the center of gravity (CG) of the aircraft is "0".

When an aircraft is in an equilibrium position, it can encounter an unexpected force. The unexpected force exerts on the aircraft, and can produce a pitch moment. This unexpected pitch moment is known as a longitudinal disturbance, and the aircraft will be in an unexpected pitching motion. The aircraft is longitudinally stable, if the aircraft is able to produce a restoring pitching moment to return to the equilibrium position: *an aircraft must have an inherent tendency to return to the pitch equilibrium after a disturbance.*

The aerodynamic forces generated on the wings, tail plane, and fuselage of an aircraft can all produce pitching moments. The wings and tail plane are all aerofoils, which generate lift and drag, when the aircraft travels in air. The magnitudes of moments produced by those forces depend on the distances from the forces to the CG. Therefore, it is clear firstly that the location of CG affects longitudinal stability. CG should not be too far back along the longitudinal axis of an aircraft, because it may enhance the disturbance and reduce the capability for the tail plane to produce sufficient restoring pitching moments to stabilize aircraft longitudinally.

Secondly, the forces generated on the mainplane are the main contributors to the pitching moment (M_m) about CG, since these forces are most significant on the aircraft and, as we discussed previously, the magnitude and location of the forces change with AoA and airspeed.

Thirdly, the tail plane, which is designed to produce aerodynamic forces by deflecting the elevator, produces a moment (M_δ) to provide the aircraft a desired pitch moment for climbing and descending and to neutralize any excessive pitching moments from other parts of the aircraft.

Lastly, fuselage can produce a pitch moment (M_α), which will affect the longitudinal stability. The pitch moment produced by fuselage changes with AoA.

The criterion of longitudinal stability is that the restoring moment by the tail plane must remain greater than any unstable moment from the mainplane. An aircraft is longitudinally stable if it can achieve that total pitch moment $M = M_m + M_\delta + M_\alpha = 0$ in level flight.

Effects on Longitudinal Stability

Different components of an aircraft are all designed for their specific functions in terms of producing aerodynamic forces and carrying out assigned tasks. However, not every component is longitudinally stable. Here we analyze the stability of main features of aircraft.

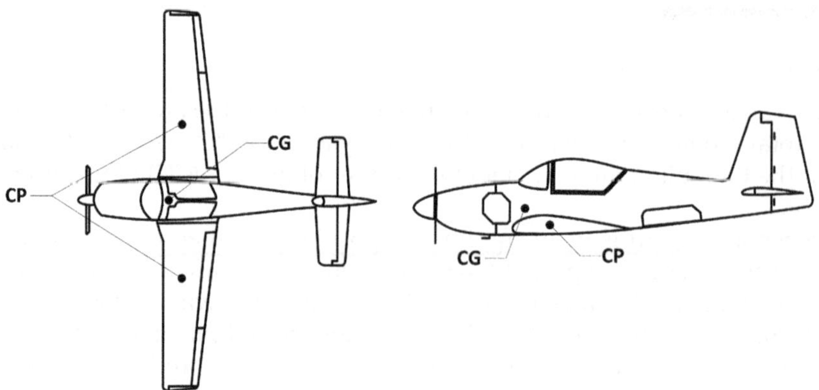

FIGURE 5.10
The relative position of CP to CG: view from above and view from side.

Wings

A wing is longitudinally stable if the CP on the wing is situated behind CG of the aircraft. For example, the wing's AoA increases, if the aircraft is a nose-up disturbance, and the wing will produce more lift. This additional lift will produce a restoring nose-down moment if CP is behind CG. This restoring nose-down pitch moment overcomes the disturbance for the aircraft to return to the equilibrium. The additional lift will produce an additional nose-up moment if CP sits in front CG. The additional nose-up moment will enhance the nose-up disturbance to push the aircraft more away from its equilibrium. So CP over a wing sitting behind CG is a stable feature. Figure 5.10 shows the different CP positions to CG.

The sweep-back wing design promotes this longitudinal stable feature. The location of their effective CP is rearward more for sweep-back wings than that of straight wings. So sweep-back wings ensure that CP is behind CG, which will ensure that the wing produces the restoring moment when there is in a pitch disturbance from its equilibrium. A heavily swept-back wing can act like a tail plane.

Center of Gravity (CG)

Longitudinal stability is about the pitching moment about CG, so the position of CG will affect the longitudinal stability, which can be influenced by the pilot of the aircraft and the operational requirements. CG depends on its own weight, fuel carried, and the payload and the load distribution. The pilot will calculate the location of its CG before flight.

A forward CG position provides a long leverage to the tail plane to produce restoring moments and ensures that CG sits before CP on the wings longitudinally. This is a stable configuration because wings can always

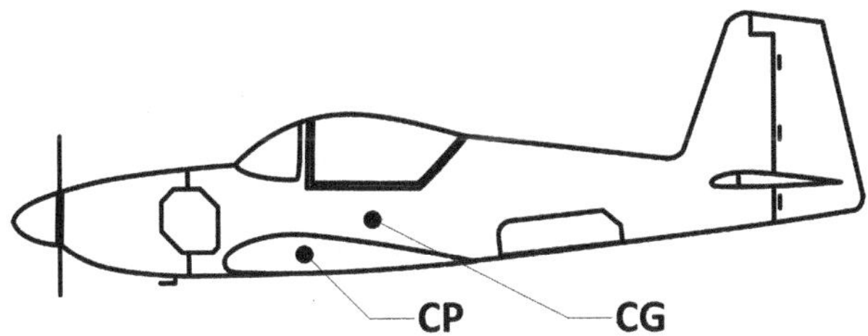

FIGURE 5.11
Unstable feature: CG behind CP.

produce a restoring moment to a pitching disturbance. However, too far forward CG will make the flight feel "nose-heavy". A CG position that is too far rearward results a short leverage to the tail plane to reduce the restoring moment produced by the tail plane. Of course, if CG is too far rearward, this is an unstable feature because it makes CG sit behind CP, as shown in Figure 5.11. Therefore, as a general practice, CG should be set in a forward position within the longitudinal limits of the aircraft.

Tail Plane

The criterion mentioned earlier shows the dominant role of tail plane in longitudinal stability.

The tail plane should be able to produce a pitch moment: $|\text{tail plane } L_{tail} \times y_{t\_CG}| > |\text{wing } L \times y_{CP\_CG}|$, where y_{t_CG} is the distance from the force L_{tail} on tail plane to CG; $y_{\_CG}$ is the distance from L, i.e. CP, on wing to CG, shown in Figure 5.12.

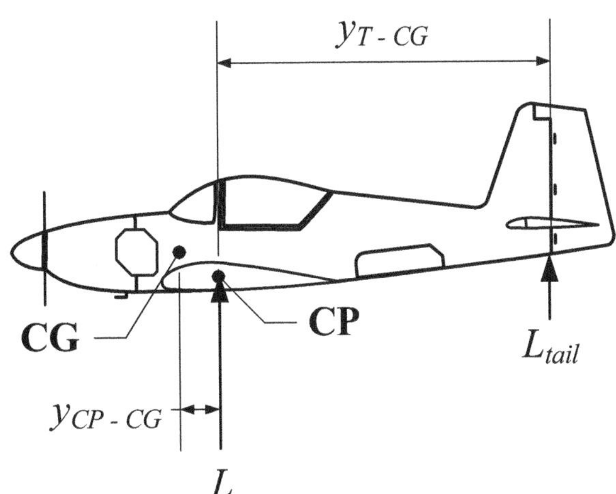

FIGURE 5.12
Aerodynamic forces on mainplane and tail plane.

The tail plane deflects in different directions to generate either "+" lift or "−" lift to balance the "nose-up" or "nose-down" moment from the wings and other parts of the airplane. When the aircraft is in nose-up disturbance, AoA increases and the tail plane produces more "+" lift, which produces a nose-down restoring moment; when the aircraft is in a nose-down disturbance, AoA decreases and the tail plane produces less "+" lift, which increases the nose-up pitch moment to restore longitudinal equilibrium.

Fuselage

Fuselage in airflow can be treated as a long solid cylinder or a large "cigar" in airflow. Local pressure varies around the body, as shown in Figure 5.13. Generally, the total force generated by integrating the pressure over the surface of the object is "0" but the sum of the moments produced by the pressure distribution is not "0".

The relative pressure distribution in Figure 5.13 is diagonally symmetric. When the AoA is positive, the total force on the first half body is an upward force, "+", if it is integrated from the leading edge to the middle of the body, while the total force on the second half body is a downward force, "−", if it is integrated from the middle to the trailing edge. These two forces have the same magnitude but act in opposite directions at different locations along the body as shown in Figure 5.13. Therefore, these two forces act like a force couple. The pitch moment produced by the force couple increases with AoA.

The fuselage will produce more nose-up pitch moment when it is in a nose-up disturbance and less nose-up moment or nose-down pitch moment if it is in a nose-down disturbance. So the fuselage can enhance the pitch disturbance and is not longitudinally stable.

Longitudinal Dihedral

The incident angle the chord of the tail plane can be different from the incident angle of the chord of main planes. (Note: the tail plane is not the T-tail,

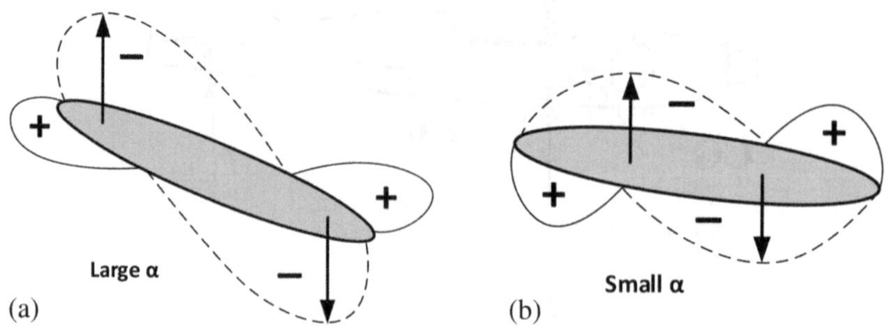

FIGURE 5.13
Pressure distribution around a "cigar-shape" body in airflow.

which is over the rudder.) For example, the incident angle of the main plane could be 3°, while the incident angle of the tail plane might be set as 1°. There is 2° difference between them. This difference is called *longitudinal dihedral*. Longitudinal dihedral can cause the airflow to form a smaller effective AoA to the wing than the actual setting of AoA, so it can decrease the magnitude of disturbance.

In normal cruise, the tail plane needs to produce a small amount of downwash force for a "nose-up" moment to keep the plane in equilibrium. The longitudinal dihedral will help the downwash on the tail plane, which can operate effectively.

Longitudinal Stability Diagram

Figure 5.14 is the longitudinal stability diagram, which is a schematic coordinate system: horizontal coordinate is AoA, and the vertical coordinate is C_{mCG}, the coefficient of pitch moment about CG. A positive C_{mCG} means a nose-up pitch moment, while a negative C_{mCG} means a nose-down pitch moment. The diagram shows the changes of coefficient of pitch moment about CG with AoA to demonstrate the longitudinal stability of a particular component of aircraft. The solid line, sitting mostly in the upper-left quadrant to the lower-right quadrant, shows the stable feature, while the dashed line, sitting in the lower-left quadrant to the upper-right quadrant, indicates the unstable feature.

The dashed line in Figure 5.14 is the unstable feature of fuselage. The dashed line indicates that C_{mCG} is positive (nose-up moment), when AoA is positive; and C_{mCG} is negative (nose-down moment), when AoA is negative. The dashed line shows an unstable feature in pitch of the fuselage: it

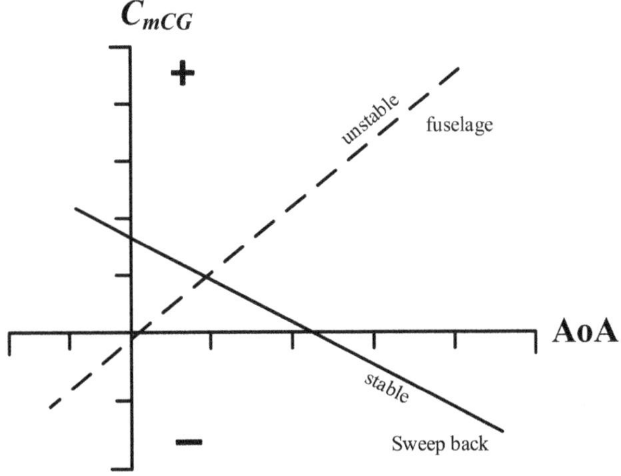

FIGURE 5.14
Longitudinal stability diagram.

produces nose-up moment when the disturbance is nose-up, and produces nose-down moment when the disturbance is nose-down. The fuselage is longitudinally unstable.

The solid line in Figure 5.14 is the stability feature of a sweep-back wing. The solid line indicates that C_{mCG} is negative (nose-down moment) when AoA is positive and C_{mCG} is positive (nose-up moment) when AoA is negative. It shows a stable feature in pitch of sweep-back wing: it produces a nose-down pitch moment when the disturbance is nose-up; and it produces a nose-up pitch moment when the disturbance is nose-down. A sweep-back wing is longitudinally stable.

We can use this diagram to analyze the longitudinal stability of other parts of aircraft, for example, the relative position of CP to CG, tail plane as was discussed earlier.

Lateral Stability

When an aircraft is in an equilibrium position and an unexpected force is exerted on the aircraft, which produces a rolling moment about the longitudinal axis, the aircraft will be in a lateral disturbance in roll, rolling away from the equilibrium position. It is said to be laterally stable if the aircraft is able to produce a restoring rolling moment to return to the equilibrium position. This is also known as "keel" effect, like the function of a keel on a ship or boat in water.

When an aircraft has disturbance in roll, it leans away from the vertical position/equilibrium position, as shown in Figure 5.15. The lift, L, and the weight of the aircraft, W, do not act in the same direction, and the (vector) sum of these two forces is F_Y, which directs to one side of the aircraft. Force, F_Y, causes the aircraft to move in its direction. This movement is *sideslip*.

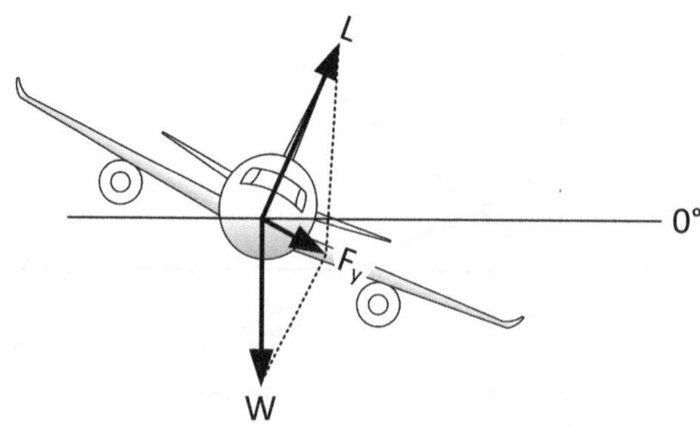

FIGURE 5.15
Schematic diagram of aircraft in roll disturbance.

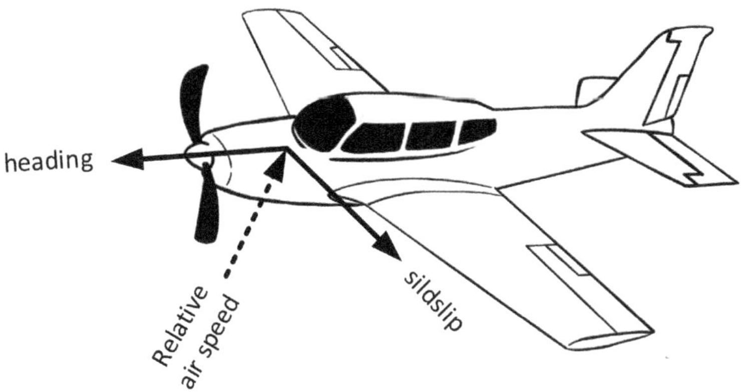

FIGURE 5.16
Sideslip and sideslip angle when aircraft is in roll disturbance.

When the aircraft is in sideslip, direction of the relative airflow changes, and is no longer in line with the aircraft's heading shown in Figure 5.16. The angle formed between the heading of aircraft and the relative airflow is *sideslip angle*. The sideslip angle is positive, "+", when the relative airflow comes from the right of heading; otherwise, it would be negative, "–". The rolling moment is positive "+", if it rolls to the right – right wing down; the rolling moment is negative, "–", if it rolls to the left – left wing down.

Several features of an aircraft are designed to provide it with positive lateral stability to ensure that the aircraft rolls back as soon as a roll disturbance occurs.

Factors on Lateral Stability

Lateral Dihedral

The wings are in the inclination from its horizontal position. The upward inclination is called *dihedral* and has a positive dihedral angle, γ, while the downward inclination is called *anhedral* and has a negative dihedral angle, as shown in Figure 5.17.

Dihedral wing can produce restoring rolling moment when aircraft is in a disturbance in roll. For example, when an aircraft with dihedral wing is in a disturbance of left roll, i.e. the left wing is down and it sideslips to the left. Its sideslip angle is negative. It means that the relative airflow (RA: or relative airspeed) comes from the left of heading, then the RA forms a greater AoA to the left wing than the AoA it forms to the upper wing – the right wing. As the result of different AoA, the left wing will produce more lift that the right wing does, as shown in Figure 5.18 (The arrows represent the lift produced by each wing). In turn, the lift on the left wing generates greater moment of right roll than the left-roll moment by the lift on the right (upper) wing.

The greater right-roll moment assists the aircraft to roll back to its original equilibrium position.

Anhedral wing has the opposite effect: the AoA of the lower wing to RA is less than that of upper wing. As the result, it produces the rolling moment in the same direction as the disturbance. This encourages the disturbance; however, this feature can be used to prevent an unstable dynamic oscillation – Dutch roll, which will be discussed later.

Shielding

The fuselage shields the trailing (upper wing) wing from the relative airflow when an aircraft is in a rolling disturbance. So the lower wing will produce more lift, then greater rolling moment to raise the lower wing to return back to its equilibrium. A similar effect occurs on the tail plane. A half of the tail plane lowers toward to the RA in a rolling disturbance, and the upper half of the tail plane is shielded from RA by the end of fuselage or rudder. The lower

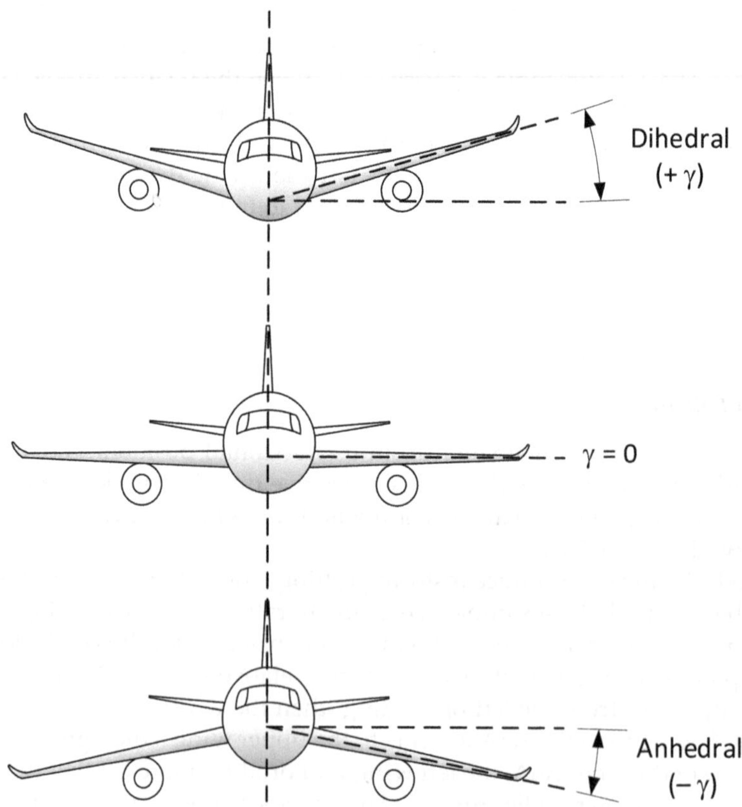

FIGURE 5.17
Dihedral and anhedral wings.

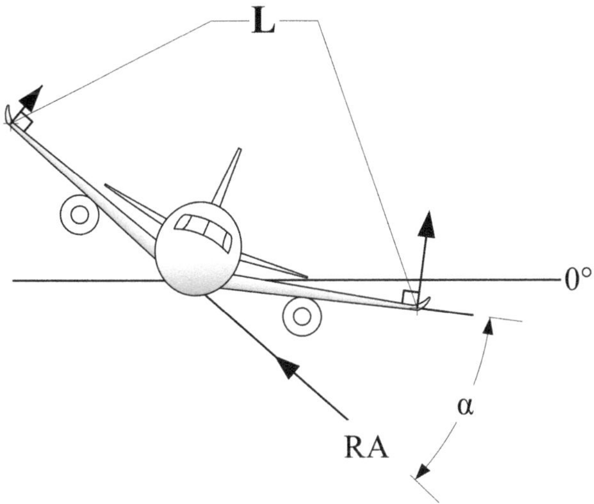

FIGURE 5.18
Dihedral wing in roll disturbance.

part of the tail plane will produce more lift than the upper part does, so the unequal forces on the tail plane will produce a restoring rolling moment.

Wing Position (Vertical)

High position wing, shown in Figure 5.19 (a): The wing is above the center of gravity of aircraft. The fuselage shields the upper wing in a rolling disturbance from RA, so the lower wing produces more lift than the upper wing dose, then, the lift on the lower wing produces higher restoring moment. When the aircraft with high position wings in a rolling disturbance, the CG swings up away from the vertical position of the longitudinal axis, as shown in Figure 5.19 (b). There is a distance between the weight W and the axis, so the weight generates rolling moment to move CG down back to its equilibrium. Therefore, the high position wing is a stable feature for lateral stability.

Low position wing, shown in Figure 5.19 (c): The wing is under the fuselage and CG of the aircraft. There is less shielding effect to the upper wing when the aircraft with low position wings in a rolling disturbance, and the CG swings up away from the vertical position of the longitudinal axis, as shown in Figure 5.19 (d). The weight W can generate the rolling moment to move CG down to be away more from its original equilibrium position.

Fin Area

Fin area produces drag to sideslip. This drag on the fin area is located above the longitudinal axis of the fuselage, producing a restoring moment. It is a positive feature to lateral stability.

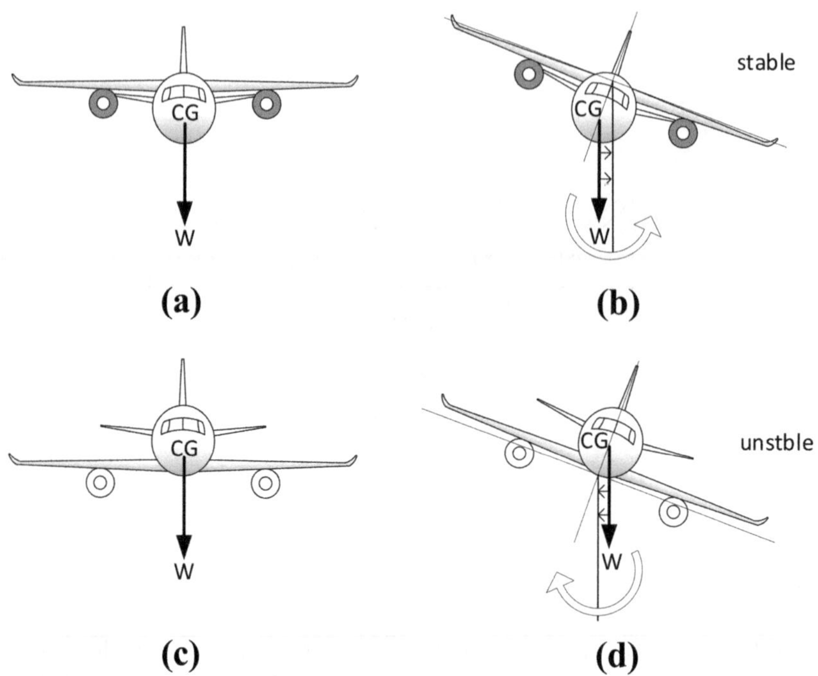

FIGURE 5.19
(a) High wing; (b) High wing in roll disturbance; (c) Low wing; (d) Low wing in roll disturbance.

Sweep-Back Wings

An aircraft with sweep-back wings in Figure 5.20 (a) is in equilibrium, and it is in a left-rolling disturbance in Figure 5.20 (b). An aircraft with sweep-back wings presents a greater effective wingspan and higher aspect ratio (AR) of its lower wing to the relative airflow (RA) than those of the upper wing, while it is in a rolling disturbance, as shown in Figure 5.20 (b). So the lower wing will produce higher lift due to its greater AR than that of upper wing. Furthermore, the lower wing produces a restoring moment in a rolling disturbance. A sweep-back wing is a stable feature in lateral stability.

Lateral Stability Diagram

Figure 5.21 is the lateral stability diagram, which is a schematic coordinate system: horizontal coordinate is sideslip angle, β, and the vertical coordinate is C_R, the coefficient of roll moment about the longitudinal axis of aircraft. A positive β means that sideslip to the right, and the RA is from the right side of heading; while a negative β means that sideslip to the left, and the RA is from the left side of heading. The sideslip angle is negative, for the aircraft in sideslip in Figure 5.20. A positive C_R means a right-roll moment, while a negative C_R means a left-roll moment.

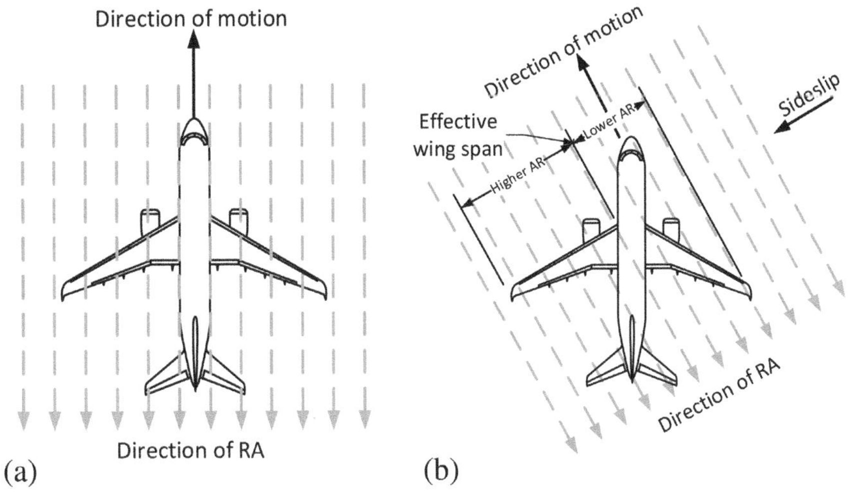

FIGURE 5.20
(a) Sweep-back wing in equilibrium; (b) Sweep-back wing in roll disturbance.

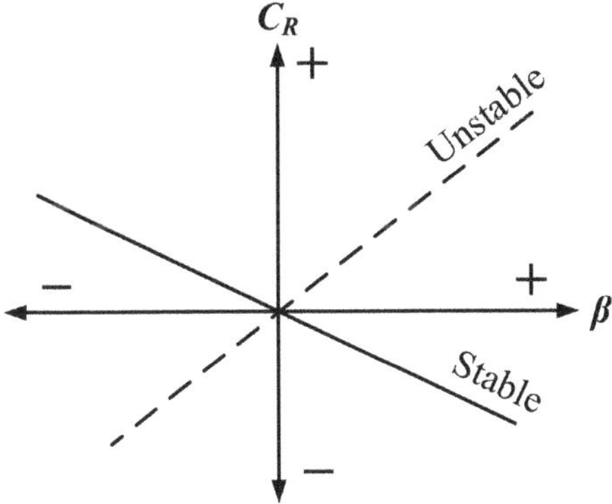

FIGURE 5.21
Lateral stability diagram.

The diagram shows the changes of coefficient of roll moment with the sideslip angle affected by a feature of an aircraft, for example, dihedral, or, anhedral wings. The solid line, sitting in the upper-left quadrant to the lower-right quadrant, shows a stable feature, while the dashed line, sitting in the lower-left quadrant to the upper-right quadrant, indicates an unstable feature.

To examine the effect of dihedral wing, for example, on the lateral stability: when an aircraft is in left-roll disturbance, and it sideslips to the left, and the left wing is lower than the right wing. The sideslip forms a negative sideslip angle β^{-}. The lower dihedral wing form a greater AoA to RA than the upper wing does. So the left wing produces more lift, and the extra lift produces a positive moment C_R^+ (in the upper-left quadrant of the diagram), i.e., the restoring moment, for the aircraft to roll back to the right so it returns to its equilibrium position. Similarly, the lower dihedral wing will produce a left roll, C_R^-, if the aircraft is in a right sideslip, β^+ (in the down-right quadrant of the diagram). So the effect of dihedral wing in this diagram is along the stable solid line.

To examine the feature of anhedral with the same method as used earlier, it can be shown that the effect of anhedral will follow the unstable dashed line in the lateral stability diagram.

Restoring/correction can only take place after a sideslip when the relative airflow forms a sideslip angle with the heading of an aircraft.

Directional Stability

Yaw is the rotating motion about the vertical axis of aircraft. The vertical axis usually passes through the CG of the aircraft. When an aircraft is in an equilibrium position and an unexpected force is exerted on the aircraft, which produces a directional moment, the aircraft will be in an unexpected yawing motion. It is said to be directionally stable if the aircraft is able to produce a restoring yawing moment to return to the equilibrium. This is known as a "weathercock effect" or "turning into the wind."

When an aircraft has disturbance in yaw, it turns away from the relative airflow RA as shown in Figure 5.22. The angle between RA and the heading of aircraft is called sideslip angle, β. If the air flow is from the right side of

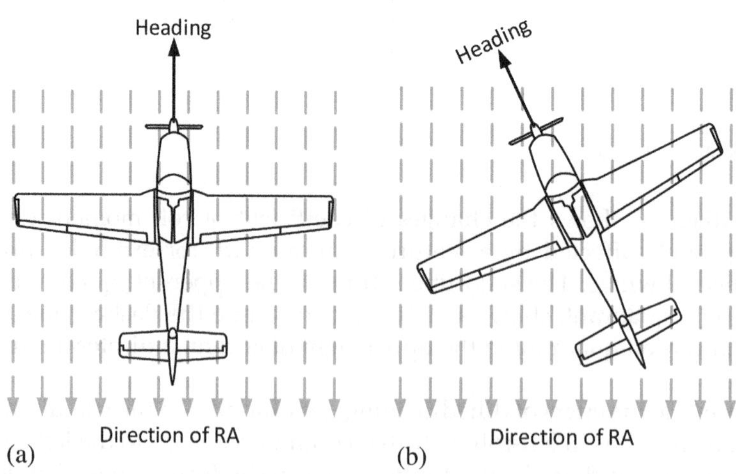

FIGURE 5.22
(a) Aircraft in equilibrium; (b) Aircraft in yaw disturbance.

the heading, the sideslip angle is positive "+", as shown in Figure 5.22 (b), otherwise, the sideslip angle is negative "–". The yawing moment is positive "+", if it makes the aircraft yaw to the right; the yawing moment is negative "–" if it makes the aircraft yaw to the left.

Factors on Directional Stability

Some features of an aircraft contribute positively to directional stability to ensure that the aircraft restores its equilibrium when it is in a yaw disturbance.

Fin

When a fin of aircraft is in a yawing disturbance, there is an angle of attack δ formed between RA and the fin, and then "lift" – the side-force, *F*, will be produced by the fin, shown in Figure 5.23. This side-force on the fin produces restoring yawing moment about CG of the aircraft, which turn the nose back into the "wind", RA. This is the "weathercock" effect by fin (on the tail plane).

There are many designs to improve the effectiveness of the fins, for example, dorsal fin and sweep-back fin, as seen in Figure 5.24. A dorsal fin is lower in height and wider at the root when compared to a slim-tall fin with

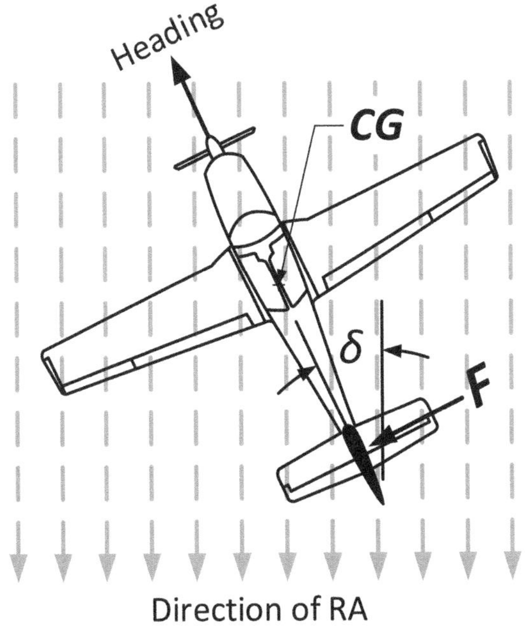

FIGURE 5.23
Force on fin in yaw disturbance.

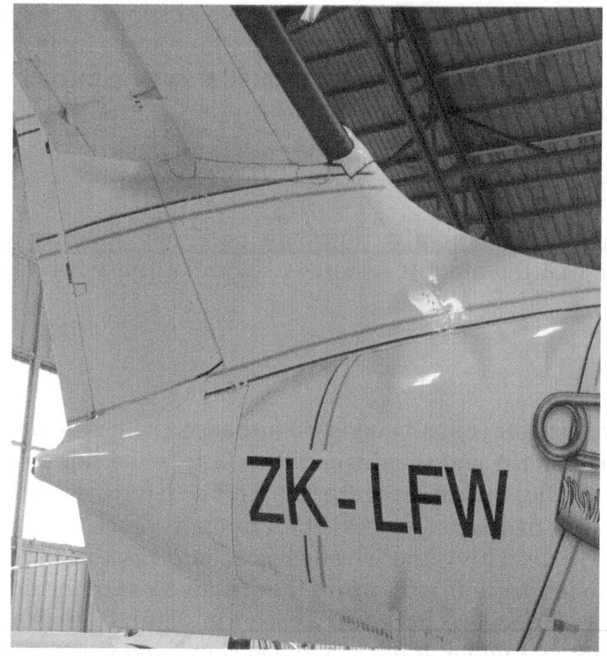

FIGURE 5.24
Dorsal sweep-back fin (Photo by author).

the same area. The shape of a dorsal fin can reduce its parasite drag because the dorsal fin root can act as a fairing to reduce airflow interference, and the lower height can reduce the boundary layer separation. Therefore, the whole area of dorsal fin can produce the "side-force" more effectively.

Similarly, a sweep-back fin can increases stall angle, so it can produce a restoring moment more effectively to overcome a relatively yaw large disturbance. A sweep-back fin can also increase the critical Mach number M_{crit} (which will be discussed in later chapters), and the fin can work effectively in the aircraft at high cruise airspeeds.

Side/"Keel" Area

The side (keel) area of aircraft is the area viewed from the side of fuselage. Drag will be produced in the side area facing to the relative airflow, when the aircraft is in a yawing disturbance. The larger side area is, the greater the drag it will produce. This drag on side area behind the CG will generate the moment to yaw the aircraft back into the "wind", i.e. RA, while the drag on the side area before the CG will generate the moment to yaw more away from the "wind". A restoring yaw moment will result if the side area behind CG is greater than that before the CG, when the aircraft is in a yaw disturbance.

Position of CG

The forward position of CG along the fuselage increases the side area was discussed earlier. The leverage of the drag on the aft side area to CG in order to increase the restoring yawing moment has to be within specific limits to move CG forward.

Sweep-Back Wing

The outer wing of an aircraft moves toward the relative airflow (RA) and the inner wing moves away from RA when the aircraft is in yaw. The outer wing travels faster than the inner wing does in yaw. The outer wing of an aircraft with sweep-back wings presents a greater effective wingspan and higher aspect ratio to RA than the inner wing, as shown in Figure 5.25. The outer wing will produce more drag, which generates the moment to yaw the aircraft back into the "wind", i.e. RA, while the inner wing will produce less drag, which generates the moment to yaw away from the "wind". Overall, sweep-back wings are a positive feature to the directional stability.

Directional Stability Diagram

Figure 5.26 is the directional stability diagram, which is a schematic coordinate system: horizontal coordinate is sideslip angle, β, and the vertical

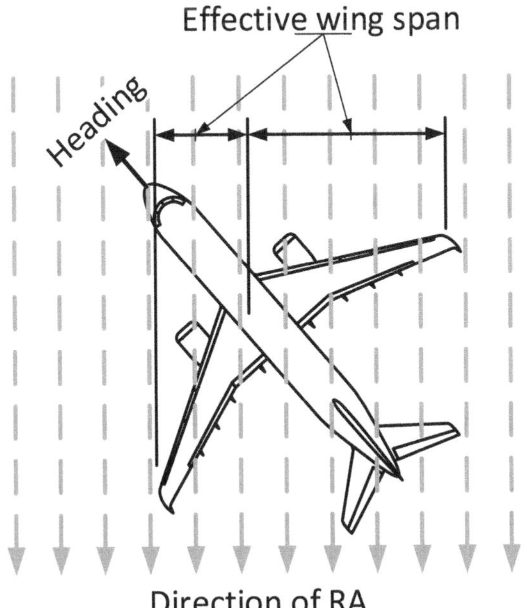

FIGURE 5.25
Sweep-back wing in yaw disturbance.

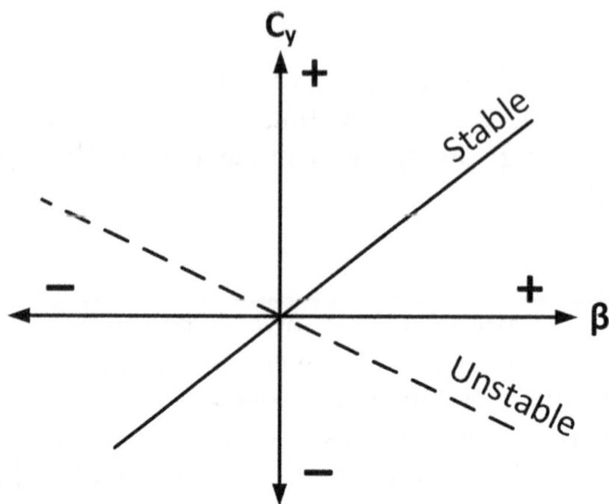

FIGURE 5.26
Directional stability diagram

coordinate is C_Y, the coefficient of yaw moment about the vertical axis through CG. A positive β means that sideslip to the left, and the RA is from the left side of heading; while a negative β means that sideslip to the right, and the RA is from the left side of heading. A positive C_Y means a right-yaw moment, while a negative C_Y means a left-yaw moment.

The diagram shows the changes of coefficient of yaw moment with the sideslip angle affected by a feature of aircraft, for example, fin, or, sweep-back wings. The solid line shows a stable feature, while the dashed line indicates an unstable feature.

To examine the effect of fin, for example, on the directional stability: when an aircraft is in a left-yaw disturbance, and it forms a positive sideslip angle, β^+, as shown in Figure 5.23. The fin forms an AoA to RA in the disturbance, and produces a side-force F heading to the left. This force will generates a right-yaw moment C_Y^+ (in the upper-right quadrant) to yaw the aircraft to the right back to its equilibrium position. Similarly, the fin will produce a left yaw, C_R^-, if the aircraft is in a yaw disturbance to the right, and forms a negative sideslip angle, β^- (in the lower-left quadrant of the diagram). So the effect of fin on directional stability is along the solid stable line in the directional stability diagram.

Analyzing the effect of sweep-back wings using the same method, we find that the sweep-back wing will produce a positive moment C_Y^+ if the aircraft is in a left-yaw disturbance. For a left-yaw disturbance, the sideslip angle is positive β_Y^+. The sweep-back wing will produce a negative moment C_Y^-, if the aircraft is in a right-yaw disturbance, i.e. the sideslip angle is negative β_Y^-. Sweep-back wings are a stable feature and follow the solid line.

We also can show the effect of the side area before and after CG in this diagram. If the side area before is greater than that after CG, we find that the side area will produce a positive moment C_Y^+ when the aircraft is in a right-yaw disturbance in a negative β_Y^-. This falls into the up-left quadrant in Figure 5.26, seen in a part of the dashed line. This kind of side area is a directionally unstable feature.

Lateral and Directional Stability (Dynamic)

It is not appropriate to discuss lateral stability only, or to discuss directional stability only to understand the stability of an aircraft, because lateral and directional stability affects each other. They are closely related.

We know the fact that roll about the longitudinal axis causes yaw about the directional axis when aircraft in air. When an aircraft is in roll, the sum of the vectors of lift and weight drives the aircraft to sideslip. It make the relative airflow, RA, come from the side of lower wing; then the aircraft will yaw into the direction of RA because of the inherent "weathercock" effect. On other hand, yaw causes roll. When an aircraft yaws, the outer wing travels faster than the inner wing does. The outer wing produces more lift than the inner wing does when the aircraft is in yaw. So a moment about the longitudinal axis is produced due to the difference in lift on the two wings to put the aircraft in roll.

Therefore, when an aircraft is in a disturbance in roll, it will experience disturbance in yaw as well, and vice versa. The aircraft will be in oscillation due to the interacting effect between these two rotating motions. Lateral stability and directional stability of an aircraft should be matched well. Ill-matched lateral and directional stability can lead to unstable oscillations.

Spiral Instability

The combination of very strong directional stability and weak lateral stability could cause *spiral instability*, which happens if the plane is in a disturbance of roll. For example, an aircraft is in a left-roll disturbance, the relative airflow (RA) comes from the left of heading because of the left sideslip. Then the strong directional stability would turn the aircraft into the "wind" immediately – left yaw before the weak lateral stability try to roll it back to the right. This left yaw causes the outer right wing travel fast to produce more lift to roll the aircraft to the left in the same direction as the disturbance, and then the strong directional stability acts again, and the left yaw and left roll occur alternately further and deeper. It is hard to restore the equilibrium from this unstable oscillation, and it can, even, turn the aircraft into a dangerous "graveyard dive".

Dutch Roll

The combination of very strong lateral stability and weak directional stability can cause an unstable oscillation called *Dutch roll*, which happens if the plane is in a disturbance of yaw. For example, an aircraft is in a left-yaw disturbance, and its heading yaws away from the relative airflow (RA). The outer right wing travels faster than the inner wing, and produces more lift to create a left-roll disturbance. The strong lateral stability will "correct" this disturbance to roll the aircraft to the right quickly before the directional stability produces an effective restoring moment. Then the directional stability and the right roll causes a right yaw, which will lead to another over-reacted left roll by the strong lateral stability, and the oscillation of yawing to one direction and rolling to the other direction at the same time continues. Dutch roll is a complex unstable oscillation. A Dutch roll with much more pronounced yawing oscillation is called *Snaking*.

Design Requirements for Lateral and Directional Stability

As we discussed earlier, it is not wise for an aircraft to have either a strong lateral or directional stability. Lateral stability and directional stability should be correctly matched. If an even match cannot be achieved, it is preferred that the lateral stability is little stronger than the directional stability is. Directional stability can be enhanced when the aircraft is in used, for example, adjusting the pro/aft position of CG by organizing payloads, and a relatively strong moment of yaw can be generated by a rudder.

Spiral instability and Dutch roll occur due to the mismatch between lateral and directional stability. Spiral can be controlled if it is not too "deep"/severe, and aircraft can tolerate some spiral instability. It is not easy to overcome/control Dutch roll, and intensified Dutch roll might cause damage to the aircraft, so Dutch roll should be dampened by design to prevent any structural damage.

Longitudinal Dynamic Stability

Disturbance oscillations can happen in pitching. In a pitch oscillation, pitch moment alternates between positive + and negative –, and oscillates about the lateral axis through CG. The amplitude of the disturbance oscillation can be dampened down with time, and gradually return to its equilibrium position. There are two modes of the longitudinal dynamic oscillation: long period mode and short period mode.

Phugoid Mode

The long period mode of longitudinal dynamic oscillation is called *Phugiod mode*. It is usually poorly dampened. The period of the oscillation is relatively long, about 20 to 60 seconds.

In this oscillation, the angle of attack along the flight path is approximately constant, and the flight path is like a sine wave going up-and-down. The magnitude of pitch moment, the altitude, and the airspeed can vary widely. In a disturbance oscillation, when aircraft reaches a higher altitude its airspeed decreases; when it reaches a lower altitude its airspeed increases. In terms of motion dynamics, it can be treated as the energy of the aircraft exchanges between potential energy and kinetic energy in its flight path.

The oscillation of this mode is "slow", not "violent", so sometimes it is not noticed by the pilot. The energy of the oscillation can eventually be dissipated, and the plane would be back to its equilibrium position.

Short Period Mode

The period of this mode of oscillation is relatively short, about 0.3 to 1.5 seconds long. In a short period oscillation, the airspeed of aircraft is approximately constant, but the angle of attack along its flight path varies in this oscillation.

The oscillation of this mode is usually well-dampened by design. The oscillation can be caused by the pitch moment from elevator flapping. To hold the control (stick-fix) at its neutral position, or release the control (stick-free) can make the plane recover from this oscillation.

Exercises

1. Does the center of pressure change with airspeed and AoA?
2. Does fuselage produce restoring moment in rolling? Is it a stable feature to lateral stability?
3. Does airspeed affect pitch moment about the AC?
4. Does AoA affect the coefficient of pitch moment about the AC?
5. By using the diagram of yawing moment coefficients, explain the feature of sweep-back wing on directional stability.
6. By using the diagram of rolling moment coefficients, explain why an airplane is laterally stable if its rolling moment coefficient is positive when the sideslip angle β is negative
7. Find the aerodynamic center and the CP of RAF 15 aerofoil. The chord is 1.8 m. Its aerofoil data are in the table below, and its $C_{M,AC} = -0.06$. Use the pitching moment about the TRAILING edge at AoA $= 4°$ ($C_L = 0.46$, $C_D = 0.02$, and $C_{M\_TE} = 0.27$, and draw a diagram to show the forces and the distances from the forces to AC, CP.)

6

Speed of Sound and Mach Number

Sound wave is a pressure wave, which is a longitudinal wave. A pressure wave travels in a medium at the speed of sound. The medium can be a solid, liquid, or gas, or a multiphase mixture. Any change in properties of the medium might lead to the change of speed of sound. Table 6.1, which is produced from the data in the *Handbook of Physics* (Benenson et al., 2002), shows the values of speed of sound in different media. Speed of sound depends on the property of the material. The data displayed in Table 6.1 were obtained under specific condition of temperature. For example, the speed of sound in air is different when air temperature is different. The speed of sound in air is the focus of this chapter.

A disturbance in air property (not heat transfer) at a location of an air system will propagate through the air at the speed of sound in all directions. For example, a piston presses air in a cylinder: The air pressure increases everywhere inside the cylinder when the piston is pushed. In this case, the speed of the piston's move is much lower than the speed of sound, so the wave of pressure change propagates throughout the cylinder before the next move of the piston, and the increase of air pressure inside the cylinder is approximately uniform.

Speed of Sound in Air

A sound wave is a pressure wave (a pressure disturbance), and a pressure wave propagates at speed of sound. In a gas system, the density, or temperature, or both density and temperature would change if the pressure of the system changed. To establish the function between the speed of sound and air property, we need to set up a flow system, which can just simply display the change of air pressure, density and temperature when the disturbance passes through the system.

The system we use here is a one-dimensional uniform air system initially, then, a pressure disturbance is produced, and this disturbance wave propagates at the speed of sound within this system. Assume that an observer sits on this wave, and travels together with the wave. To the observer, the air flows steadily from one side of the wave to the other side at the speed of wave, v, and the speed and air property are assumed to change once it passes

TABLE 6.1

Speed of sound in various materials (20°C).

Material	Speed (ms^{-1})
Air	343
Air (0°C)	331
Carbon dioxide	258
Water	1480
Sea water	1470
Kerosene	1451
Aluminum	5200
Titanium	6070

through the wave, as shown in Figure 6.1. There isn't either heat exchange nor mass loss when the airflow passes the wave, so it is an adiabatic process, and the mass flow rate through the wave can be expressed by the parameters of the airflow at both sides of the wave. Therefore, we can use the Continuity equation (2.1) and the Euler equation (2.7) for this one-dimensional steady flow.

The pressure, density, and temperature of the air are p, ρ, and T. Air pressure becomes $p+\delta p$ after the airflow passes the pressure disturbance (pressure wave). δp is the change in air pressure, and then, the air density, temperature, and its speed all assumed to have changed accordingly after the pressure wave, and become $\rho + \delta\rho$, $T+\delta T$, and $v+\delta v$.

Apply the Continuity equation (2.1):

$$\rho v A = (\rho + \delta\rho)(v + \delta v)A \tag{6.1}$$

where A is the area of the wave front, which is a constant to the air flow of both sides. The Equation (6.1) becomes:

$$\rho v = (\rho + \delta\rho)(v + \delta v) = C \tag{6.2}$$

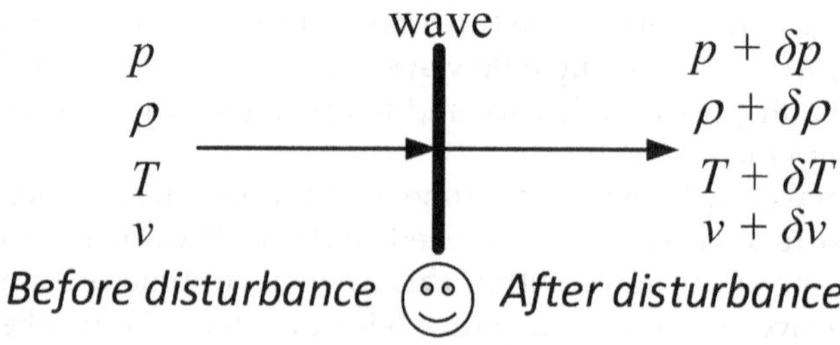

FIGURE 6.1
A pressure wave with an observer.

Differentiate (6.2) and rearrange:

$$\rho dv + v d\rho = 0; \Rightarrow \frac{\text{v}}{\rho} = -\frac{dv}{d\rho} \tag{6.3}$$

There is no change in elevation when the steady airflow passes the wave, so it is a level flow. Apply the Euler equation (2.7):

$$\frac{1}{\rho}\frac{dp}{ds} + v\frac{dv}{ds} = 0 \Rightarrow dv = -\frac{1}{\rho v}dp \tag{6.4}$$

Substitute Equation (6.4) into (6.3):

$$\frac{v}{\rho} = \frac{1}{\rho v}\frac{dp}{d\rho}; \Rightarrow v^2 = \frac{dp}{d\rho}; \tag{6.5}$$

This speed is the speed of the pressure wave, and it is *Speed of Sound,* and denoted by *a*.

$$a^2 = \frac{dp}{d\rho}, \text{ and } a = \sqrt{\frac{dp}{d\rho}} \tag{6.6}$$

Air is treated as an ideal gas, and it is in an adiabatic process when air passes the pressure disturbance wave. From Equation (2.29), ideal gas in adiabatic process, we obtain $\frac{p}{\rho^\gamma} = C$. Differentiate it:

$$\frac{1}{\rho^\gamma}dp - \gamma\frac{p}{\rho^{\gamma+1}}d\rho = 0; \Rightarrow \frac{dp}{d\rho} = \gamma\frac{p}{\rho} \tag{6.7}$$

Substitute Equation (6.7) into (6.6), and apply the Ideal Gas Law (2.20):

$$a = \sqrt{\frac{dp}{d\rho}} = \sqrt{\gamma\frac{p}{\rho}} = \sqrt{\gamma R_M T}; \quad \text{Unit}: [\text{ms}^{-1}] \tag{6.8}$$

where T is the absolute temperature of the gas in K; R_M is gas constant; and γ is the adiabatic index of the gas.

Equation (6.8) shows the speed of sound in an ideal gas, and shows that the *speed of sound is a function of the temperature of the ideal gas, and temperature only*. In air transport, the unit of speed, knots [kt], is commonly used. For air, $\gamma = 1.4$, Equation (6.8) can be converted to:

$$a = 39\sqrt{T}; \quad \text{Unit}: [\text{kt}] \tag{6.9}$$

where T is in K.

Example 6.1

1. Find the speed of sound in air a, if the air temperature is $T = -50°C$.
2. If the condition of air: $p = 1.02 \times 10^5$ Pa, and $\rho = 1.225$ kg/m^3, calculate a.

SOLUTION

1. Use Equation (6.8): $a = \sqrt{\gamma R_M T} = \sqrt{1.4 \times 287 \times (273 - 50)} = 299.33\ (\text{ms}^{-1})$

   Or, use Equation (6.9): $a = 39\sqrt{T} = 33\sqrt{(273 - 50)} = 582.39\ (\text{kt})$

   The speed of sound is 299.33 ms^{-1}, or 582.39 kt.

2. Use Equation (6.6): $a = \sqrt{\gamma \dfrac{p}{\rho}} = \sqrt{1.4 \times \dfrac{1.02 \times 10^5}{1.225}} = 341.43\ (\text{ms}^{-1})$

   The speed of sound is 341.43 ms^{-1}.

Mach Number

When air particles travel at a speed, which is greater than the speed of sound, any change in air property would not be able to propagate throughout the whole airflow system. The ratio of the air speed v to the local speed of sound a indicates how close the air speed to the speed of sound. This ratio is the *Mach number*. It is dimensionless, and is denoted as M:

$$M = \frac{v}{a} \tag{6.10}$$

The Mach number is named after Ernst Mach (1838–1916), an Austrian physicist and philosopher who specialized in optics and who was a pioneer in study of supersonic fluid mechanics and shockwaves.

The Mach number, M, is an important parameter to indicate the characteristics of a high-speed compressible fluid flow:

- Airflow is *subsonic* when $M < 1$. The air particles travel at a speed that is smaller than the local speed of sound;
- Airflow is *sonic* if $M = 1$. The air particles travel at the speed of sound;
- Airflow is *supersonic* when its $M > 1$. The air particles travel at a speed that is greater than the speed of sound.

Example 6.2

1. What is the speed of sound in air, a, under sea level conditions?
2. What is the speed of sound a, if air pressure and density are $p = 1.20 \times 10^5$ Pa and $\rho = 1.225$ kgm^{-3}, respectively?

3. An aircraft travels at an altitude where the temperature is $T = -10°C$, and its Mach number is $M = 0.9$. What is its airspeed (in ms^{-1} and in kt)?

SOLUTION

1. The temperature at sea level is 288 K. According to Equation (5.6), the speed of sound in air is $a = \sqrt{\gamma R_M T} = \sqrt{1.4 \times 287 \times 288} = 340.17\ ms^{-1}$.
2. Use IGL: $T = \frac{p\rho}{R_M} = \frac{1.20 \times 10^5 \times 1.225}{287} = 341.32$ K, and the speed of sound of the air is $a = \sqrt{\gamma R_M T} = \sqrt{1.4 \times 287 \times 341.32} = 370.32\ ms^{-1}$.
3. The temperature is $T = 273 - 10 = 263$ K. From the definition of Mach number Equation (5.8), it is shown:
   $v = Ma = 0.9 \times \sqrt{1.4 \times 287 \times 263} = 292.57\ ms^{-1}$, or $v = 0.9 \times 39\sqrt{263} = 569.22$ kt.

Mach Wave

When the relative airflow around an object, for example, an aircraft, is supersonic, the relative air speed v is greater than the speed of sound, $v > a$. The object is the source of the change of air property, for example, air pressure. The pressure wave propagates at the speed of sound, so the pressure change always sits behind the object, which is demonstrated in Figure 6.2.

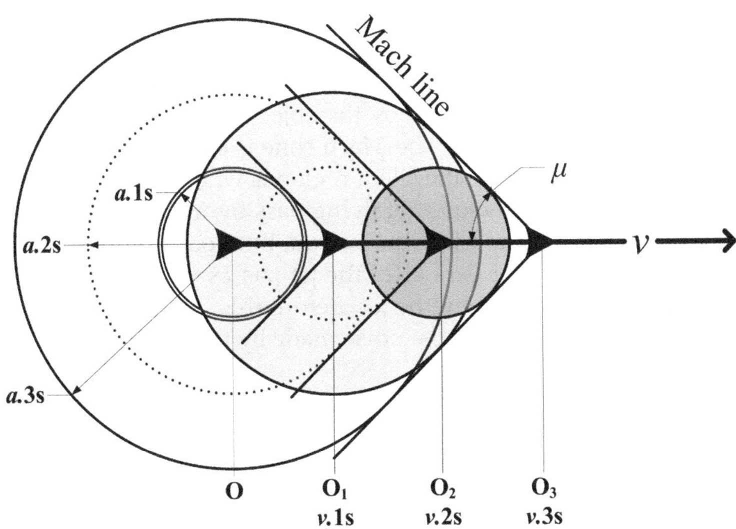

FIGURE 6.2
Propagation of disturbance in supersonic flow.

In Figure 6.2, the object starts from point "O" at the speed $v > a$. The object reaches "O_1", "O_2", and "O_3", after traveling 1 second, 2 seconds, and 3 seconds, respectively. The pressure disturbance wave propagates at speed of sound, a, from the object spherically.

> The wave fronts are represented by the circles in Figure 6.2. The pressure wave from "O" reaches the double-bold circle, after the object travels for 1s, and arrives at "O_1". The distance between point O and point O_1 is $v \cdot 1$s, and the radius of the double-bold circle is $a \cdot 1$s.
>
> The circles with dashed-line are the pressure wave fronts from "O" and "O_1" after the object travels for 2s, and arrives at "O_2". The distance between point O and point O_2 is $v \cdot 2$s, and the radius of the dashed circle from "O" is $a \cdot 2$s, and the radius of the dashed circle from "O_1" is $a \cdot 1$s.
>
> The circles with a solid line are the pressure wave fronts from "O", "O_1", and "O_2" after the object travels for 3s. The distance between point O and point O_3 is $v \cdot 3$s. The radii of the solid-line-circles from "O", "O_1", and "O_2" are $a \cdot 3$s, $a \cdot 2$s, and $a \cdot 1$s, respectively.

A common tangent line can be drawn from the object to the circles, which represent the pressure wave fronts at any moment, and the common tangent line is *Mach wave* or *Mach line*. In three-dimensional space, the Mach lines form a surface, which is in conical shape, called a *Mach cone*. The half apex-angle, μ, of the Mach cone is given:

$$\sin\mu = \frac{at}{vt} = \frac{a}{v} = \frac{1}{M}, \quad \text{and} \quad \mu = \sin^{-1}\left(\frac{1}{M}\right) \tag{6.11}$$

a is the local speed of sound, v is the speed the object travels at, and t is travel time. μ is called the *Mach angle.*

The pressure disturbance caused by the object can only propagate within the Mach cone, so the outside of the Mach cone is called the "silent region". One of the ways to experience the silent region is when you are in a military air show and you see a supersonic jet flying pass-by. You would not hear the jet coming, even it were right in front of you, but you could suddenly hear a loud sound strike in your ears after the jet passes you, because you now are in the Mach cone caused by the jet. Before this, you were in the "silent region" and you could not hear the noise made by the jet.

Shockwaves

Shockwaves will be formed when airflow becomes supersonic if there are compressive pressure disturbances. A shockwave is an accumulated wave front, which is very thin, less than 1 mm. The details of shockwave formation will be discussed in Chapter 7. Air pressure, density, and temperature will be all increased suddenly after a shockwave.

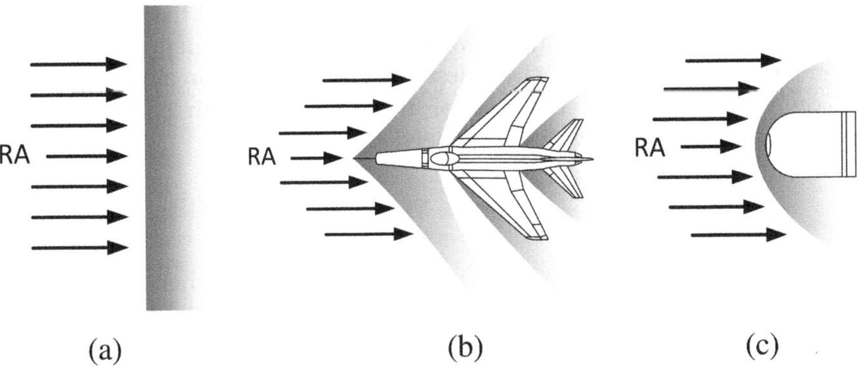

FIGURE 6.3
(a) Normal shockwave; (b) Oblique shockwave; (c) Bow shockwave.

There are three types of shockwaves:

Normal shockwave – the wave front is perpendicular to the direction of the relative airflow, RA, as shown in Figure 6.3 (a);

Oblique shockwave – the wave front is not perpendicular to the direction of the relative airflow, RA, as shown in Figure 6.3 (b);

Bow shockwave – a combination of normal shockwave and oblique shockwave, as shown in Figure 6.3 (c).

When a shockwave propagates in the air, and there are sudden changes in air property, in particular, the air pressure. The sudden change of pressure strikes every object along its path including the ears of observers. Observers' ears will experience the sudden pressure change as a high-pitched bang – a high-intensity sound like an explosion. This sudden strike is called *sonic boom*.

We will hear a "sonic boom" when watching a supersonic aircraft flying pass-by, and when whipping crack.

Special Mach Numbers

An aircraft travels at a speed in air, which is usually called true airspeed (TAS). This true air speed is also called *free-stream airspeed*, v_{fs}.

Free-stream Mach number M_{fs}: The ratio of the true airspeed, or free-stream airspeed to speed of sound in the free-stream, a_{fs}:

$$M_{fs} = \frac{TAS}{a_{fs}} = \frac{v_{fs}}{a_{fs}} \tag{6.12}$$

However, the speed of airflow around the aircraft usually is not always the same as the free-stream airspeed, and it varies from location to location. For example, the local airspeed over a part of upper surface of the aerofoil usually is greater than the free-stream airspeed, but the local airspeed under the lower surface of the aerofoil can be less than the free-stream airspeed.

Local Mach number M_L: The ratio of a local airspeed (air particle speed) around an aircraft, v_{local}, to the local speed of sound, a_{local}:

$$M_L = \frac{v_{local}}{a_{local}} \tag{6.13}$$

Critical Mach number M_{crit}: The free-stream Mach number when a local Mach number reaches "1". This Mach number marks the first local air (particles) speed becomes sonic.

Example 6.3

The true airspeed of an aircraft is 180 ms^{-1}, at sea level conditions. The highest airflow speed reaches 230 ms^{-1} at the location of the maximum camber of its aerofoil. Calculate the free-stream *Mach* number, M_{fs}, and the local *Mach* number, M_L of the airflow at the most camber point. Is the M_{fs} the critical *Mach* number?

SOLUTION

The free-stream temperature is under sea level conditions: $T = 288$ K. Use Equation (6.8) to calculate the free-stream speed of sound, for air, $\gamma = 1.4$ and $R_M = 287$ J/kg.K:

$$a_{fs} = \sqrt{\gamma R_M T_{fs}} = \sqrt{1.4 \times 287 \times 288} = 340.17 \text{ ms}^{-1}$$

So the free-stream Mach number by using the definition (6.12): $M_{fs} = \frac{TAS}{a_{fs}} = \frac{180}{340.17} = 0.53$.

To calculate the local Mach number M_L at the location where $v = 230$ ms^{-1}, we need the local temperature. According to the Energy equation (2.38), as discussed in Chapter 2, the local temperature will decrease while the airspeed increases:

$$C_p \times 288 + \frac{1}{2} \times 180^2 = C_p \times T_{\text{local}} + \frac{1}{2} \times 230^2,$$

and Equation (2.39) shows, $C_p = \frac{\gamma R_M}{\gamma - 1} = \frac{1.4 \times 287}{1.4 - 1} = 1004.5 \text{ J(kg.K)}^{-1}$,

so $T_{\text{local}} = 288 + \frac{1}{2 \times 1004.5} \times (180^2 - 230^2) = 277.8 \text{ (K)}$

Use Equation (6.13): $M_L = \frac{v_{\text{local}}}{a_{\text{local}}} = \frac{230}{\sqrt{1.4 \times 287 \times 277.8}} \doteq 0.69$.

The free-stream Mach number is 0.53, while the highest local Mach number is 0.69, so the free-stream Mach number is not the critical Mach number.

Ranges of Flights

The free-stream Mach number of an aircraft is used to classify the flights.

Subsonic flights: The free-stream *Mach* number of aircraft is less than its critical *Mach* number, $M_{fs} < M_{crit}$. The airflow around the subsonic aircraft is always subsonic, i.e., the local air flow speed around a subsonic aircraft is always less than the local speed of sound. The airflow can be treated as incompressible if the true airspeed of the aircraft is less than 250 kt, low subsonic. Otherwise, the airflow around the subsonic is compressible.

Transonic flights: The free-stream Mach number of aircraft is greater than its critical Mach number, and less than, approximately, 1.2, $M_{\text{crit}} < M_{fs} < 1.2$. The airflow around transonic aircraft can be subsonic, as well supersonic, even when the free-stream Mach number is less than 1. The air definitely is compressible, and shockwaves may be formed on aerofoils and on other parts of the aircraft body.

Supersonic flights: The free-stream Mach number of aircraft is greater than 1.2, $M_{fs} > 1.2$. This is also called hypersonic if the free-stream Mach number of any traveling object is greater than 5 or 6. Airflow around the supersonic aircraft is supersonic in general, except the airflow behind a normal shockwave, and within boundary layers. The air is highly compressible, and the kinetic heating is a significant concern due to the speed change of airflow around supersonic aircraft.

Mach Number Measurement

The Mach number is monitored in flight. It requires the information of airspeed (TAS) and the temperature at the altitude the aircraft is at in order to obtain the Mach number of the aircraft. A normal shockwave will be formed at the leading edge of the Pitot tube or a probe that measures total pressure of the airflow if the airflow speed exceeds the speed of sound. The method to obtain the Mach number is different for aircraft in the different flight range due to the aerodynamic characteristics at that range.

Low Subsonic Flight

For the flights in the low subsonic range, the airflow around the aircraft can be treated as incompressible fluid flow. A Machmeter obtains the Mach number by using the ratio of dynamic pressure, p_d, to static pressure p_s.

The dynamic pressure = the total pressure p_t (stagnation pressure p_o) – static pressure p_s, according to Bernoulli's equation (2.14), the ratio used in the Machmeter is:

$$\frac{p_t - p_s}{p_s} = \frac{\frac{1}{2}\rho v^2}{p_s} = \frac{1}{2}\frac{v^2}{\frac{p_s}{\rho}}. \tag{6.14}$$

Applying the Ideal Gas Law, Equation (6.14) becomes:

$$\frac{p_t - p_s}{p_s} = \frac{1}{2}\frac{v^2}{\frac{p_s}{\rho}} = \frac{\gamma}{2}\frac{v^2}{\gamma R_M T}$$

and then use Equation (6.8):

$$\frac{\gamma}{2}\frac{v^2}{\gamma R_M T} = \frac{\gamma}{2}\frac{v^2}{a^2} \propto \frac{v^2}{a^2} = M^2 \tag{6.15}$$

So the Machmeter can obtain the Mach number by using:

$$M \propto \sqrt{\frac{p_t - p_s}{p_s}} \tag{6.16}$$

High Subsonic and Low Transonic Flight

Air cannot be treated as incompressible when airspeed is greater than 250 kt. The ambient air density at the location, where the static pressure is measured, is different from the stagnation density. The expression (6.16) will not be correct for compressible airflow.

A Pitot tube in a compressible airflow can still produce the pressure difference between the stagnation air pressure, p_o, and static air pressure, p. The airflow inside the Pitot tube should satisfy the Energy equation (2.37), when the airflow reaches the stagnation, which is in an adiabatic (isentropic) process. Rearrange the Energy equation (2.37) and expression of C_p (2.39):

$$C_p T + \frac{1}{2}v^2 = C_p T_o \Rightarrow 1 + \frac{1}{2}\frac{(\gamma - 1)v^2}{\gamma R_M T} = \frac{T_o}{T} \tag{6.17}$$

By using Equation (6.8) for the speed of sound and Equation (2.29) of gas property in adiabatic process, Equation (6.17) can be reorganized as:

$$\frac{1}{2}\frac{(\gamma - 1)v^2}{\gamma R_M T} = \frac{(\gamma - 1)}{2}\frac{v^2}{a^2} = \frac{T_o}{T} - 1;$$

then,

$$\frac{v^2}{a^2} = \frac{2}{(\gamma-1)}\left(\frac{T_o}{T}-1\right) = \frac{2}{(\gamma-1)}\left(\left(\frac{p_o}{p}\right)^{\frac{\gamma-1}{\gamma}}-1\right) \tag{6.18}$$

For air, $\gamma=1.4$, so Equation (6.18) is:

$$\frac{v^2}{a^2} = M^2 = 5\left(\left(\frac{p_o}{p}\right)^{\frac{2}{7}}-1\right) \tag{6.19}$$

where p is static pressure, $p=p_s$, and p_o is stagnation pressure, which is the total pressure p_t from Pitot tube. So the Machmeter can obtain the Mach number by:

$$M = \sqrt{5\left(\left(\frac{p_t}{p_s}\right)^{\frac{2}{7}}-1\right)} \tag{6.20}$$

Supersonic Flight

A normal shockwave will be formed in front of a Pitot tube, which measures the total, or stagnation pressure, when it is in a supersonic airflow. The energy of the airflow will be decreased after it passes through the shockwave, and the process is not reversible (not isentropic), so Equation (6.20) cannot be used in the supersonic airflow.

Figure 6.4 shows a bow shockwave formed in front of an operating Pitot tube. The opening of the Pitot tube has to be behind the normal part of the bow shockwave, so the stagnation (or total) pressure sensed by the Pitot tube is the stagnation pressure after the shockwave.

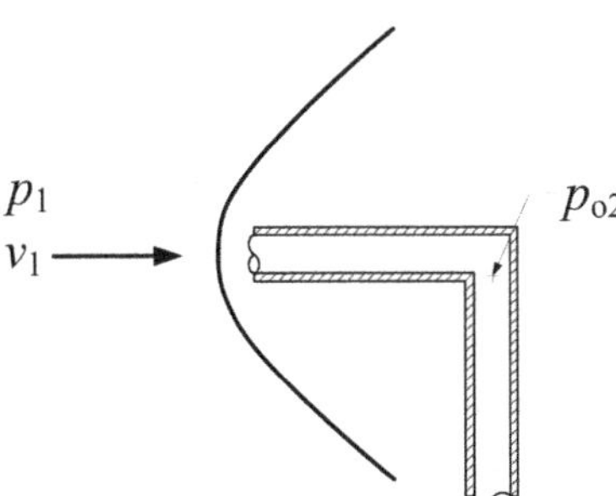

FIGURE 6.4
Schematic diagram of Pitot tube in supersonic flow.

The static pressure of the free-stream and the stagnation pressure behind the normal shockwave in front Pitot tube can be measured. Chapter 7 will introduce the relationships of air property before and after a normal shockwave. Using the relationships of air property before and after a normal shockwave, and the relationship of static pressure, speed, and the stagnation pressure, Equation (6.21) gives the expression of the ratio of the total pressure p_{o2} after the shockwave and the free-stream static pressure p_1 in terms of free-stream (incoming) Mach number M:

$$\frac{p_{o2}}{p_1} = \frac{\left(\frac{\gamma+1}{2}M^2\right)^{\frac{\gamma}{\gamma-1}}}{\left[\frac{2\gamma M^2-(\gamma-1)}{\gamma+1}\right]^{\frac{1}{\gamma-1}}} \tag{6.21}$$

This equation is called the *Rayleigh supersonic Pitot tube equation.*

Rearrange (6.21) for air, $\gamma = 1.4$, and it can be written as:

$$M = k\sqrt{\left(\frac{p_{o2}}{p_1}\right)^{\frac{2}{7}}(7M^2-1)^{\frac{5}{7}}} \tag{6.22}$$

where k is constant; p_{o2} is the total pressure after the normal shockwave in front of the Pitot tube, and p_1 is the free-stream static pressure. The pressures can be measured on board of a plane, and the incoming Mach number – free-stream Mach number can be calculated by Equation (6.22) using an iterating method. The calculation should be carried out by the on-board computer.

Mach Number Applications

The display of flight data in a cockpit draws usually the information of a specific parameter from different sensing sources. For example, TAS can be multidisplayed with the signals from different Pitot tubes, and altitude (ALT) can be found from ALT sensor, and as well as from GPS information. Sometimes, the values of a specific parameter from different sources disagree with each other. As we have discussed earlier, the Mach number is the ratio of TAS and speed of sound; speed of sound is a function of temperature; and air temperature is a linear function of ALT in troposphere according to ISA (International Standard Atmosphere). Pilots can, then, perform cross-checking to eliminate the incorrect one with the principle of the Mach number. The following examples explain the method of cross-checking.

Example 6.4

An aircraft travels at 300 kt, (TAS), and the on-board Machmeter shows that the Mach number is 0.5. Estimate the altitude (ALT) of the aircraft.

SOLUTION

From the definition of Mach number, $M = 0.5 = \frac{TAS}{a} = \frac{300}{a}$;

The local speed of sound is $a = \frac{300}{0.5} = 600$ (kt).

Use Equation (6.9), the local temperature is $T = \left(\frac{600}{39}\right)^2 = 236.7$ (K).

Apply ISA model:

$$T = T_{\text{sealevel}} - 1.98\,(\text{K/kft})\text{ALT};\ \text{where ALT is in kft, and } T \text{ is in } K. \qquad (6.23)$$

At sea level $T = 288$ K, so $\text{ALT} = \frac{(288 - 236.7)}{1.98} \doteq 25.9$ (kft)

The estimated altitude is ALT = 25,900 ft. The pilot can compare the estimated ALT with the displayed ALT in the cockpit.

Example 6.5

The ALT of an aircraft is 20,000 ft, and its airspeed is 260 kt. The on-board Machmeter shows $M = 0.44$. Estimate the Mach number to check.

SOLUTION

Use ISA Equation (6.23); the local temperature is $T = 288 - 1.98\,(\text{K/kft}) \times 20 = 248.4$ (K).

Use Equation (6.9) to calculate local speed of sound: $a = 39\sqrt{248.4} = 614.7$ (kt).

Then $M = \frac{260}{614.7} = 0.423$, which is less than 0.44.

The estimated Mach number is lower than the displayed Mach number. The pilot should check if the Machmeter works correctly. If it does, the disagreement in Mach number might indicate that the local temperature is lower than the expected temperature calculated by ISA. The pilot might need to check the weather report to confirm if the aircraft has encountered a cold weather patch.

Monitoring the Mach number in flight, when aircraft is in climbing, descending, or banking, can raise the pilot's awareness to a change in TAS. For example, if an aircraft's Mach number is constant while it is climbing, the pilot should realize that the aircraft's TAS is decreasing because the ambient temperature decreases, and the local speed of sound decreases. So the constant Mach number tells the pilot that the TAS is decreasing. The pilot should ensure that the aircraft does not decrease its TAS too much, in particular, if it is close to its low speed limit.

Exercises

1. The Mach number of an aircraft is $M=0.9$, and the local temperature is $T=-10°C$. What is its airspeed?
2. Estimate TAS if an aircraft is at ALT = 9500 m and its Mach number M is 0.5.
3. An aircraft is in flight, and its TAS = 220 m/s. The ambient temperature $T=245$ K. A local airspeed over an aerofoil reaches 314 m/s. What is M_{fs}? Has the M_{fs} exceeded its M_{crit}?
4. Can you calculate the local speed of sound if you know ALT only? Is it true that the speed of sound definitely increases if both the air pressure and density are increasing?
5. Give flight situations in which both an aircraft's IAS (indicated airspeed) and Mach number will not change.

7

Compressible Air Flow

Air naturally is a compressible fluid, particularly when the airspeed is greater than 250kt. The assumption of incompressible air is simple and easy to use in the analysis of airflow in Chapter 2, but it introduces significant differences between the theoretical analysis and the reality when the speed of airflow is relatively high, close to the speed of sound, so this assumption is no longer valid. In order to understand the fundamental characteristics of compressible airflow, which suits the purpose of air transport application, this chapter considers one-dimensional airflow following its streamlines.

To analyze the relationship between the speed of airflow and the properties of air, we use a set of mathematical equations to describe the state of the air and the motion, and energy change of the airflow.

Compressible 1-D Airflow System

The equations used to analyze the airflow are called a set of control equations, which describe the nature and characteristics of the air and air movement. We consider the airflow within its flow path s with air property, p, ρ, and T, at speed v. The airflow is in steady state, i.e. $\frac{\partial}{\partial t} = 0$; there is no heat exchange to and from the airflow; and it is assumed that friction is negligible and the change in elevation of the flow is negligible as well. In this chapter the set of control equations are:

Continuity Equation

The Continuity equation describes the mass conservation of the airflow path: the air flow rate is constant:

$$\rho v A = C \tag{7.1}$$

as described by Equation (2.1), where A is the cross-section area of the flow path.

Differentiate (7.1), following total derivative form (1.12):

$$vAd\rho + \rho A dv + \rho v dA = 0$$

and divide above by $\rho v A = C$, then obtain the differential form of Continuity equation:

$$\frac{d\rho}{\rho} + \frac{dv}{v} + \frac{dA}{A} = 0 \tag{7.2}$$

Momentum (Euler) Equation

Consider the airflow is in steady state, and the change in height is negligible, the Euler equation (2.7) for 1-D fluid flow can be simplified as:

$$v\frac{dv}{ds} = \frac{1}{\rho}\frac{dp}{ds} \tag{7.3}$$

Then

$$vdv = -\frac{1}{\rho}dp \tag{7.4}$$

Energy Equation

As it is assumed that there is no heat exchange to or from the airflow. It will satisfy the Energy equation for fluid flow (2.27):

$$C_pT + \frac{v^2}{2} = C_pT_o = E \tag{7.5}$$

where T_o is the stagnation temperature; E is a constant; $C_p = \frac{\gamma R_M}{\gamma - 1}$, the specific heat of gas when pressure is constant; and $\gamma = \frac{C_p}{C_V}$, and $C_V = \frac{R_M}{\gamma - 1}$, the specific heat when the volume(density) is constant, while R_M is the gas constant. The unit of R_M, C_p, and C_V is J(kg.K)$^{-1}$.

Ideal Gas Law (State Equation of Gas)

Air is treated as an ideal gas:

$$\frac{p}{\rho} = R_MT \tag{7.6}$$

Critical Point

The Energy equation (7.5) shows that the air temperature decreases if airspeed increases and air temperature increases when airspeed decreases. The air temperature reaches its maximum when airspeed decreases to "0". In

Chapter 6, we learned that the speed of sound in air (Ideal Gas) is a function of temperature, temperature only: $a = \sqrt{\gamma R_m T}$. Using the expression speed of sound a to replace air temperature T, the Energy equation (7.5) can be rewritten as:

$$C_p \frac{\gamma R_m T}{\gamma R_m} + \frac{v^2}{2} = \frac{C_p a^2}{\gamma R_m} + \frac{v^2}{2} = E$$

Substitute $C_p = \dfrac{\gamma R_m}{\gamma - 1}$ into above:

$$\frac{a^2}{\gamma - 1} + \frac{v^2}{2} = E \tag{7.7}$$

The Energy equation (7.7) shows that the speed of sound a in the airflow changes with airspeed, v. Speed of sound reaches its maximum when airspeed decreases to "0", and vice versa. Speed of sound is a function of temperature, T, so a and v all functions of T. The speed of sound, a, changes from "0" to maximum and airspeed v will change from its maximum to "0" while the temperature changes from "0" to the stagnation temperature T_o. And, somewhere in the range of its temperature, the airspeed and the speed of sound of the airflow will be identical.

For example, if the speed of airflow, v is 250 m/s at sea level conditions, the changes of v and the speed of sound a with temperature T can be displayed in Figure 7.1, if the temperature could vary from absolute zero (which is imaginary situation) to the stagnation temperature T_o. Figure 7.1 shows that the changes of the speed of sound of the airflow, a, and the airspeed, v, with the air temperature. a is increasing (dashed line) and v is decreasing (solid line) from maximum to "0", when T changes from "0" to "T_o". These two speed lines intercept when the temperature is at approximately 266 K. It means that $v = a$, at the intercept.

The intercept is the *critical point*, at which the airspeed is the critical airspeed, v_c, and the speed of sound is critical speed of sound, a_c. $v_c = a_c$, so the local Mach number $M = 1$. The temperature at the intercept is critical temperature T_c. The critical temperature T_c and the critical speed of sound a_c can be calculated by Equation (7.7): $v = a = a_c$,

$$\frac{a^2}{\gamma - 1} + \frac{v^2}{2} = E = C_p T_o = \frac{v_c^2}{\gamma - 1} + \frac{a_c^2}{2}$$

then:

$$C_p T_o = \frac{a_c^2}{\gamma - 1} + \frac{a_c^2}{2} = \frac{\gamma + 1}{2(\gamma - 1)} a_c^2 \tag{7.8}$$

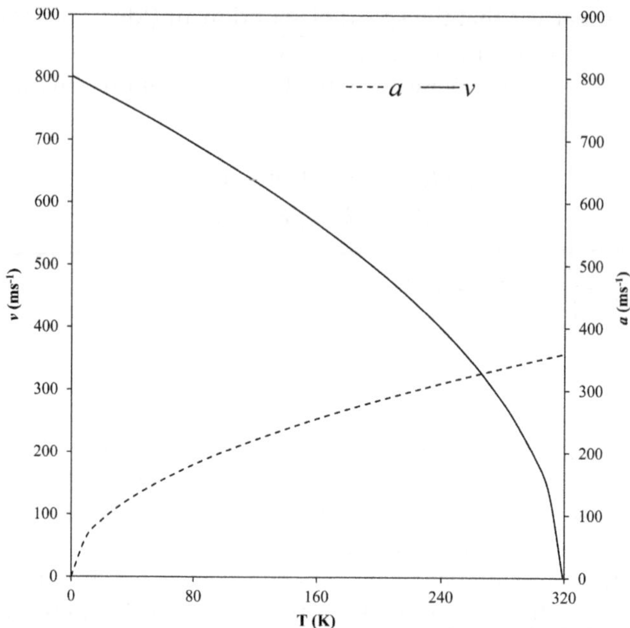

FIGURE 7.1
Change of airspeed v and speed of sound with temperature of an airflow.

Rearrange Equation (7.8) and the formula of speed of sound (6.8):

$$a_c = \sqrt{\frac{2(\gamma - 1)}{(\gamma + 1)} C_p T_0} = \sqrt{\gamma R_M T_c}. \tag{7.9}$$

Furthermore, applying the expression of C_p (2.39) to (7.9), one can obtain a formula to calculate the critical temperature T_c:

$$T_c = \frac{2}{(\gamma + 1)} T_0. \tag{7.10}$$

The critical air properties, p_c, and ρ_c, and the stagnation air properties, p_o, and ρ_o, can all be calculated by the formulae of gas properties in an adiabatic process (2.29):

$$\begin{cases} \dfrac{p_c}{p} = \left(\dfrac{T_c}{T}\right)^{\frac{\gamma}{\gamma - 1}} \\ \dfrac{\rho_c}{\rho} = \left(\dfrac{T_c}{T}\right)^{\frac{1}{\gamma - 1}} \end{cases} \tag{7.11}$$

$$\begin{cases} \dfrac{p_o}{p} = \left(\dfrac{T_o}{T}\right)^{\frac{\gamma}{\gamma-1}} \\ \dfrac{\rho_o}{\rho} = \left(\dfrac{T_o}{T}\right)^{\frac{1}{\gamma-1}} \end{cases} \tag{7.12}$$

Examples 7.1

(a) Assuming that the $M_{crit}=0.75$, what is the airspeed of the aircraft when it reaches M_{crit} at sea level? Is it true that the local airspeed at $M_L=1$ is about 340 ms^{-1}?

(b) For the same aircraft: what is the airspeed of the aircraft when it reaches M_{crit} at 20,000 ft (248 K)? What is the airspeed at $M_L=1$?

SOLUTION

(a) At sea level: $T=288$ K, the airspeed is

$$v = 0.75a = 0.75\sqrt{\gamma R_m T} = 0.75\sqrt{1.4\times 287\times 288}$$

$$= 0.75\times 340.2 = 255.15\ (\text{ms}^{-1})$$

The airspeed is 255.15 ms^{-1} when the free-stream Mach number reaches M_{crit}.
The airflow reaches its critical point when the local Mach number $M_L=1$, where $v_c=a_c$. To calculate a_c, we need to know T_o. Use the Energy equation (7.7):

$$T_o = T + \frac{1}{2C_p}v^2 = 288 + \frac{255.15^2}{2\times 1004.5} = 320.4\ (\text{K})$$

while:

$$C_p = \frac{\gamma R_m}{\gamma - 1} = \frac{1.4\times 287}{0.4} = 1004.5\ (\text{J/kgK})$$

Use Equation (7.9):

$$v_c = a_c = \sqrt{2\left(\frac{\gamma-1}{\gamma+1}\right)C_p T_o} = 327.5\ (\text{ms}^{-1})$$

So, it is not true that the airspeed at the location, where $M_L=1$, is 340 ms^{-1}, and the airspeed should be 327.5 ms^{-1}at the location.

(b) At 20,000ft: $T=248$ K, when $M_{fs}=M_{crit}=0.75$, the airspeed will be:

$$v = 0.75a = 0.75\sqrt{\gamma R_m T} = 0.75\sqrt{1.4\times 287\times 248} = 236.8\ (\text{ms}^{-1})$$

The airspeed is 236.8 ms^{-1} when it reaches the M_{crit} at 20,000 ft, and this air speed is less than that at sea level.

The stagnation temperature at 20,000ft, similarly, can be calculated by (7.7):

$$T_o = T + \frac{v^2}{2C_p} = 248 + \frac{236.8^2}{2 \times 1004.5} = 275.9\text{ K}$$

When $M_L = 1$, the airflow reaches the critical point, the critical airspeed is by Equation (7.9):

$$v_c = a_c = \sqrt{2\left(\frac{\gamma - 1}{\gamma + 1}\right) C_p T_o} = 303.9\ (\text{ms}^{-1})$$

At the location of $M_L = 1$ over the aerofoil, the local airspeed is 303.9 ms^{-1}, and it is less that the corresponding airspeed at sea level.

Speed Coefficient *M**

Speed coefficient is a ratio of airspeed, v, to its critical speed of sound, a_c, and it is denoted as $M^* = \frac{v}{a_c}$. Mach number $M = \frac{v}{a}$, as we know, v is airspeed, and a is the local speed of sound. As Figure 7.1 shows, the speed of sound changes with the airspeed if there is no heat exchange to the airflow and the airflow satisfies the Energy equation, but a_c is a constant to the airflow. M changes with v and a, while M^* only changes with v. Figure 7.2 shows the change of M and M^* with temperature in the airflow displayed in Figure 7.1.

M^* is not the Mach number, which represents the airflow status related to sonic level, but it can indicate if the airflow is subsonic or supersonic just as the Mach number M does. Let us take the Energy equation (7.8), and consider the air speed and temperature at any point along the flow stream and the speed and temperature at its critical point:

$$\frac{a^2}{\gamma - 1} + \frac{v^2}{2} = \frac{(\gamma + 1)a_c^2}{2(\gamma - 1)} \tag{7.13}$$

Divide both sides of (7.13) by v^2:

$$\frac{1}{(\gamma - 1)M^2} + \frac{1}{2} = \frac{(\gamma + 1)}{2(\gamma - 1)M^{*2}} \tag{7.14}$$

Rearrange (7.14):

$$M^* = \sqrt{\frac{(\gamma + 1)}{2}} \frac{M}{\sqrt{1 + \frac{(\gamma - 1)}{2} M^2}} \tag{7.15}$$

or

$$M = \sqrt{\frac{2}{(\gamma + 1)}} \frac{M^*}{\sqrt{1 - \frac{(\gamma - 1)}{(\gamma + 1)} M^{*2}}} \tag{7.16}$$

(7.15) and (7.16) displays the relationship of M and M^*.

When $M^*=1$, from (7.16) shown here:

$$M=\sqrt{\frac{2}{(\gamma+1)}}\frac{1}{\sqrt{1-\frac{(\gamma-1)}{(\gamma+1)}}}=\sqrt{\frac{2}{(\gamma+1)}}\frac{M^*}{\sqrt{\frac{2}{(\gamma+1)}}}=1 \quad (7.17)$$

When the speed coefficient is $M^*=1$, $M=1$, the flow is sonic.

When $M^*<1$, $\sqrt{1-\frac{(\gamma-1)}{(\gamma+1)}}<\sqrt{1-\frac{(\gamma-1)}{(\gamma+1)}M^{*2}}$.

We know that from (7.17) $\sqrt{\frac{2}{(\gamma+1)}}\frac{1}{\sqrt{1-\frac{(\gamma-1)}{(\gamma+1)}}}=1$

So $\sqrt{\frac{2}{(\gamma+1)}}\frac{1}{\sqrt{1-\frac{(\gamma-1)}{(\gamma+1)}M^{*2}}}<1$

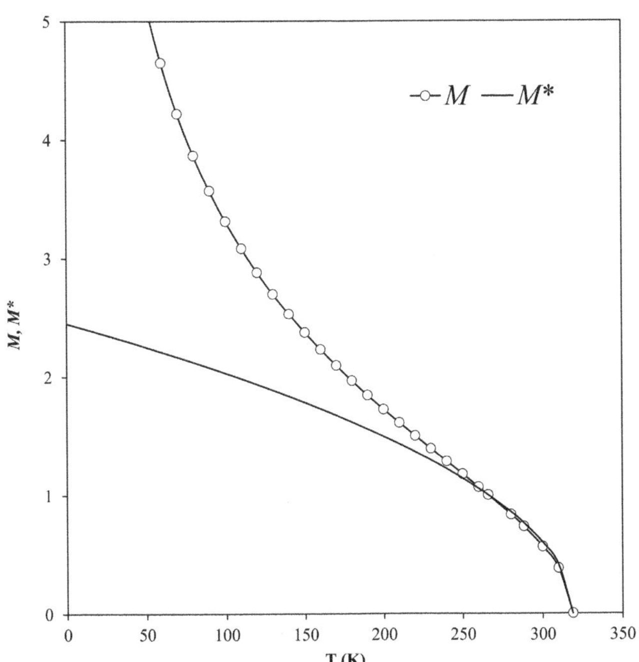

FIGURE 7.2
Change of Mach number M and speed coefficient M^* with temperature of an airflow.

which leads Equation (7.16) to

$$M=\sqrt{\frac{2}{(\gamma+1)}}\frac{M^*}{\sqrt{1-\frac{(\gamma-1)}{(\gamma+1)}M^{*2}}}<M^*.$$

When the speed coefficient is $M^*<1$, the airflow definitely is subsonic because $M<M^*<1$.

When $M^*>1$, $\sqrt{1-\frac{(\gamma-1)}{(\gamma+1)}}>\sqrt{1-\frac{(\gamma-1)}{(\gamma+1)}M^{*2}}$.

We know that from (7.17) $\sqrt{\frac{2}{(\gamma+1)}}\frac{1}{\sqrt{1-\frac{(\gamma-1)}{(\gamma+1)}}}=1$.

So $\sqrt{\frac{2}{(\gamma+1)}}\frac{1}{\sqrt{1-\frac{(\gamma-1)}{(\gamma+1)}M^{*2}}}>1$

which leads Equation (7.16) to

$$M=\sqrt{\frac{2}{(\gamma+1)}}\frac{M^*}{\sqrt{1-\frac{(\gamma-1)}{(\gamma+1)}M^{*2}}}>M^*$$

When the speed coefficient is $M^*>1$, the airflow definitely is subsonic because $M>M^*>1$.

Figure 7.3 shows the relationship between M and M^*, as discussed earlier. For subsonic airflow, M is always lower than M^*. For supersonic airflow, M is greater than M^*. When airflow reaches sonic, $M=M^*=1$.

Compressible Airflow with a Variable Area of Flow Path

Air properties will change along a flow path if the cross-section of the flow path varies. Equations (7.2) and (7.4) describe the motion of a level and steady compressible airflow, and the speed of sound in air can be expressed by Equation (6.6). Following a level steady airflow through flow path, the Continuity equation (7.2) shows that total change of its mass flowrate is "0":

$$\frac{d\rho}{\rho}+\frac{dv}{v}+\frac{dA}{A}=0$$

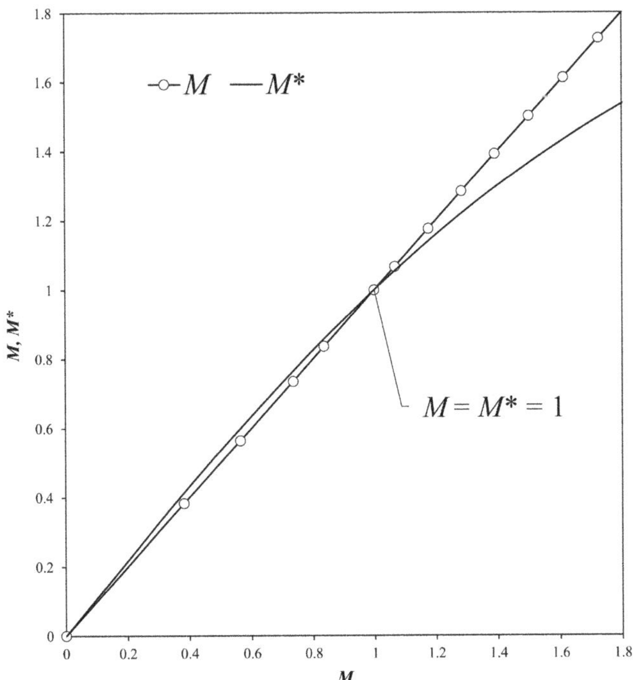

FIGURE 7.3
The relationship between Mach number M and speed coefficient M^*.

Use Equation (6.6), $a^2 = \dfrac{dp}{d\rho}$ in the Continuity equation:

$$\frac{d\rho}{\rho a^2}\frac{dp}{d\rho} + \frac{dv}{v} + \frac{dA}{A} = \frac{dp}{\rho a^2} + \frac{dv}{v} + \frac{dA}{A} = 0 \tag{7.18}$$

Substitute the simplified Euler equation (7.4), $vdv = -\dfrac{dp}{\rho}$, into (7.18):

$$\frac{-vdv}{a^2} + \frac{dv}{v} + \frac{dA}{A} = \left(1 - \frac{v^2}{a^2}\right)\frac{dv}{v} + \frac{dA}{A} = 0$$

Rearrange above, then obtain:

$$\boxed{\frac{dA}{A} = (M^2 - 1)\frac{dv}{v}} \tag{7.19}$$

Rearrange Equation (7.4):

$$dv = -\frac{dp}{v\rho} \tag{7.20}$$

Substitute Equation (7.20) into Equation (7.19):

$$\frac{dA}{A} = (M^2 - 1)\frac{-dp}{v^2\rho} \tag{7.21}$$

Apply the Ideal Gas Law (7.6) and $a^2 = \gamma R_M T$ to Equation (7.21):

$$\frac{dA}{A} = (1 - M^2)\frac{R_M T dp}{v^2 p} = (1 - M^2)\frac{a^2}{\gamma v^2}\frac{dp}{p},$$

then,

$$\boxed{\frac{dA}{A} = \frac{(1 - M^2)}{\gamma M^2}\frac{dp}{p}} \tag{7.22}$$

Equations (7.19) and (7.22) state the relationships between the changes in speed, air pressure, and the area of the flow path of a compressible airflow.

For a subsonic airflow, where $M < 1$, $(M^2 - 1)$ is negative, and $(1 - M^2)$ is positive:

> Equation (7.19) and Equation (7.22) show that the airspeed will increase, $\frac{dv}{v} > 0$, and the air pressure will decrease, $\frac{dp}{p} < 0$, when the flow path is getting narrower, $\frac{dA}{A} < 0$. In this case, the airflow is in a "convergent nozzle" as shown in Figure 7.4 (a). The local Mach number will increase with the airspeed.
>
> On the other hand, Equation (7.19) and Equation (7.22) show that the airspeed will decrease, $\frac{dv}{v} < 0$, and the air pressure will increase, $\frac{dp}{p} > 0$, when the flow path is getting wider, $\frac{dA}{A} > 0$. The airflow is, then, in a "diffusor" as shown in Figure 7.4 (b).

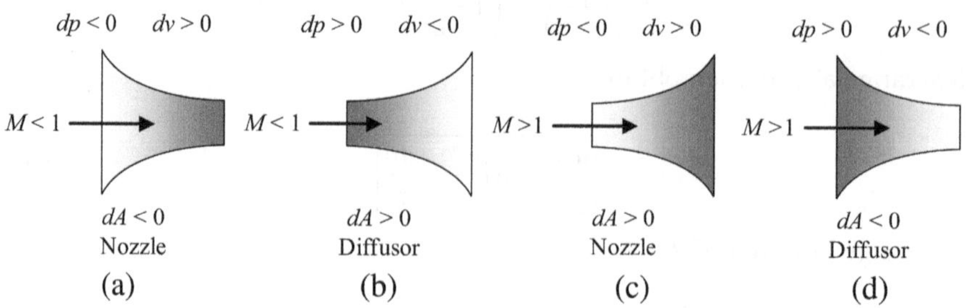

FIGURE 7.4
(a) Convergent nozzle; (b) Divergent diffusor; (c) Divergent nozzle; (d) Divergent diffusor.

For a supersonic airflow, where $M > 1$, $(M^2 - 1)$ is positive, and $(1 - M^2)$ is negative:

> Equation (7.19) and Equation (7.22) show that the airspeed will increase, $\frac{dv}{v} > 0$, and the air pressure will decrease, $\frac{dp}{p} < 0$, when the flow path is getting wider, $\frac{dA}{A} > 0$. In this case, the airflow is in a "divergent nozzle" as shown in Figure 7.4 (c).
> On the other hand, Equation (7.19) and Equation (7.22) show that the airspeed will decrease, $\frac{dv}{v} < 0$, and the air pressure will increase, $\frac{dp}{p} > 0$, when the flow path is getting narrower, $\frac{dA}{A} < 0$. The airflow is in a "diffusor" as shown in Figure 7.4 (d).

The airflow processes shown in Figure 7.4 are all reversible adiabatic processes, so the change in air property can be determined by the gas state equations for adiabatic process (2.29).

Converging-Diverging Nozzle

Combining the convergent nozzle in Figure 7.4 (a) and the divergent nozzle in Figure 7.4 (c) forms a convergent-divergent nozzle shown in Figure 7.5. This nozzle can increase an airspeed from subsonic to supersonic.

At the inlet of the converging-diverging nozzle, the airflow is subsonic, $M < 1$. The airspeed increases and air pressure decreases in the convergent section of the nozzle, as discussed earlier, and the local Mach number increases.

The area of the cross-section of this nozzle changes continuously from $\frac{dA}{A} < 0$ to $\frac{dA}{A} > 0$ from the inlet to the outlet of the nozzle. At one place along the nozzle, $\frac{dA}{A} = 0$, which means that A reaches its extreme value. This extreme value is a local minimum, that is, the throat of the nozzle. $\frac{dv}{v}$ is

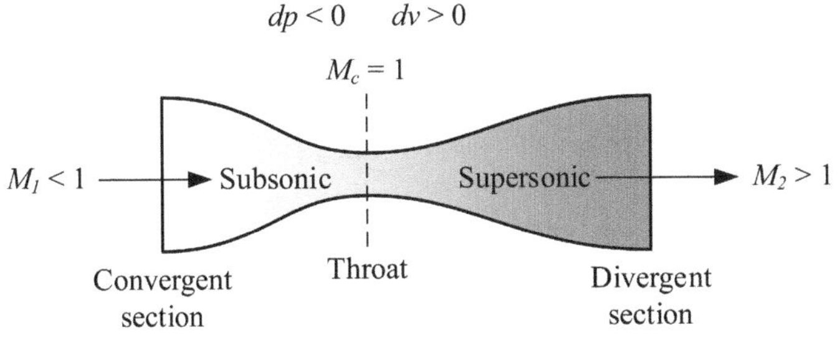

FIGURE 7.5
A convergent-divergent nozzle.

positive and the Mach number of the airflow increases in the convergent section of the nozzle. When the Mach number increases to "1", Equation (7.19) shows that $M^2-1=0$, and $dA=0$, at the throat, where the airflow is at its critical point, and $\frac{dv}{v}$ continues to be positive.

In the divergent section, from the throat to the outlet of the nozzle: from the throat Mach number starts being greater than 1, $M>1$, and $dA>0$, area increases. Airspeed, v, and air pressure, p, will continue increasing and decreasing respectively in the divergent section, according to Equations (7.19) and (7.22). Therefore, airflow is supersonic, $M>1$ at the exit of the nozzle.

The airflow in a convergent-divergent nozzle satisfies the set of control Equations (7.2), (7.3), (7.5), and (7.6), and the process within the nozzle is reversible. It can be shown by the similar analysis that the whole process of airflow in a convergent-divergent nozzle would be reversed if the airflow at the inlet were supersonic. $M=1$ at the throat, and airflow would be subsonic, $M<1$, at the outlet. Air pressure would increase throughout the whole path from inlet to outlet.

Example 7.2

Airflow enters a convergent-divergent nozzle at 200 m/s when $T_1=200°C$ and $p_1=2.02\times10^5$ Pa.

(a) Calculate the air properties at the throat, the critical point: T_c, p_c, and ρ_c.
(b) If the temperature at the exit of the nozzle is $T_2=100°C$, what is the air speed, and the Mach number at the exit?
(c) What is the ratio of areas $A_{1(in)}/A_{2(out)}$?

SOLUTION

(a) The temperature at the inlet of the nozzle is $T_1=200+273=473$ K, and airspeed $v_1=200$ ms^{-1}. Use Equation (7.10) to calculate the critical temperature, $T_c=\frac{2}{\gamma+1}T_o$. Use the Energy equation (7.7) and the expression of C_p, or $C_p=1004.5$ J(k.kg)$^{-1}$ to calculate the stagnation temperature, T_o:

$$T_c=\frac{2}{\gamma+1}T_0=\frac{2}{\gamma+1}\left(T_1+\frac{v_1^2}{2C_p}\right)$$

$$=\frac{1}{\gamma+1}\left(2T_1+\frac{(\gamma-1)}{\gamma R_M}v_1^2\right)$$

$$=\frac{1}{2.4}\left(2\times473+\frac{0.4\times200^2}{1.4\times287}\right)=411\,(\text{K})$$

Use Equation (7.11) to calculate the critical pressure, p_c and critical density ρ_c, or use the Ideal Gas Law for ρ_c once T_c and p_c are known:

$$\frac{p_c}{p_1}=\left(\frac{T_c}{T_1}\right)^{\frac{\gamma}{\gamma-1}}=\left(\frac{411}{473}\right)^{\frac{1.4}{1.4-1}}, \quad \text{and } p_c=2.02\times10^5\times\left(\frac{411}{473}\right)^{\frac{1.4}{1.4-1}}=1.235\times10^5 \text{ (pa)}$$

Use the Ideal Gas Law:

$$\rho_c=\frac{p_c}{R_M T_c}=\frac{1.235\times10^5}{287\times411}=1.047 \text{ (kgm}^{-3}\text{)}$$

So, at the throat of the nozzle, air pressure is 1.235×10^5 Pa; air density is 1.047 kgm^{-3}; and air temperature is 411 K or about 138°C.

(b) At the exit, $T_2=100°\text{C}=373$ K. Use the Energy equation (7.7) to calculate the airspeed v_2.

$$C_pT_1+\frac{v_1^2}{2}=C_pT_2+\frac{v_2^2}{2}\Rightarrow v_2^2=v_1^2+2C_p(T_1-T_2)$$

$$v_2=\sqrt{200^2+2\times1004.5\times(473-373)}=490.8 \text{ (ms}^{-1}\text{)}$$

The Mach number at the exit of the nozzle is

$$M_2=\frac{v_2}{a_2}=\frac{490.8}{\sqrt{1.4\times287\times373}}=1.27.$$

(c) Apply the Continuity equation $A_1\rho_1v_1=A_2\rho_2v_2$ to find the ratio of areas $A_{1(\text{in})}/A_{2(\text{out})}$.
The values of air density at the inlet and outlet of the nozzle are required to use the Continuity equation.
Obtain the density at the inlet, ρ_1, by using the Ideal Gas Law:

$$\rho_1=\frac{p_1}{R_MT_1}=\frac{2.02\times10^5}{287\times473}=1.488 \text{ (kgm}^{-3}\text{)}$$

Obtain the density at the outlet, ρ_2, by using the adiabatic relationship:

$$\frac{\rho_2}{\rho_1}=\left(\frac{T_2}{T_1}\right)^{\frac{1}{\gamma-1}}=\left(\frac{373}{473}\right)^{\frac{1}{1.4-1}}, \quad \text{and } \rho_2=1.488\times\left(\frac{373}{473}\right)^{\frac{1}{1.4-1}}$$

$$=0.822 \text{ (kgm}^{-3}\text{)}$$

Applying the Continuity equation at inlet and outlet, the ratio of the area at inlet to the area at outlet is:

$$\frac{A_1}{A_2}=\frac{\rho_2v_2}{\rho_1v_1}=\frac{0.822\times490.8}{1.488\times200}=1.36$$

Shockwave

The disturbance/change of air property, for example, a change in air pressure, in an airflow system propagates at speed of sound, as stated in Chapter 6. The main difference between subsonic and supersonic airflow systems is the fact that air particles travel slower than the propagation of its property change in subsonic airflow system, while air particle travels faster than the propagation of its property change in a supersonic system. For example, in a cylinder, the air pressure can increase immediately once the piston is pushed in, because the speed of sound the pressure wave travels at is much greater than the speed of the air particles inside the cylinder. The pressure is assumed to be uniform inside the cylinder at any moment of time. The same thing cannot be assumed in a supersonic airflow system. A type of pressure disturbance can cause a shockwave in a supersonic system because the air particles travel at a speed greater than speed of sound.

Formation of Shockwave

In a supersonic airflow, $v > a$, if there is a source of *compression* pressure disturbance, the source of the disturbance (the air particles) travels faster than the disturbance, or the pressure change. Because the "change" never can catch up with the air particles because air particles travel faster than speed of sound, the air property is not uniform, and the "change" is not continuous throughout the airflow system. So the disturbances from the moving the source accumulate (or build-up) to form a strong wave front, and this wave front is *shockwave*. The thickness of a shockwave is, typically, about 1 mm, very thin, and air properties change suddenly across a shockwave.

Air Properties Before and after a Normal Shockwave

Air property and airspeed change suddenly when a supersonic airflow passes through a shockwave. This process is irreversible. However, there is no loss or gain of air mass, and there is no heat exchange to or from this system in this process, so the airflow passing through a shockwave should follow the Continuity equation and the Energy equation.

Figure 7.6 shows airflow passes a normal shockwave, whose front is perpendicular to the direction of the airflow (its velocity). The airflow system we consider here is the shockwave and the airflow in the immediate vicinity before and after the shockwave. It is assumed that this system is a one-dimensional, and the viscous friction within the airflow is negligible. The air properties and the airflow speed, Mach number before and after a shockwave are p_1, v_1, ρ_1, v_1, and M_1, and p_2, v_2, ρ_2, v_2, and M_2, respectively.

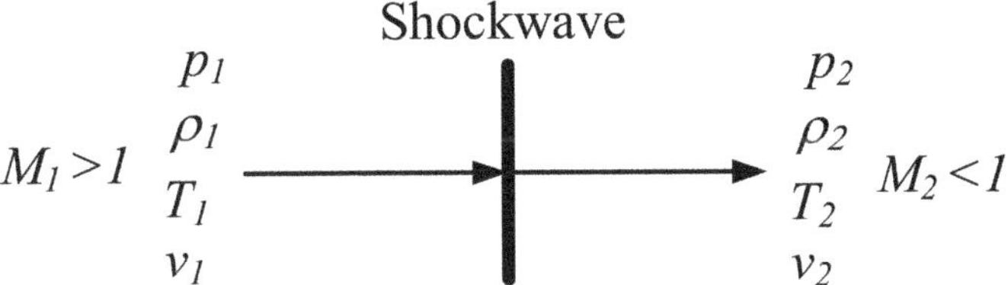

FIGURE 7.6
Airflow passes through a normal shockwave.

The mass flow rate of the airflow is constant when it passes cross the normal shockwave, so apply Continuity equation (7.1), while the area of the flow path is constant:

$$\rho_1 v_1 A = \rho_2 v_2 A \Rightarrow \rho_1 v_1 = \rho_2 v_2 \tag{7.23}$$

where A is the common area of the shockwave.

The change of momentum of the airflow before and after the shockwave can be determined by the Impulse-Momentum Theorem:

$$\sum F = \Delta(\dot{m}v), \tag{7.24}$$

where $\dot{m}$ is the mass flow rate of the air. The total force on the air particles exerted by pressure across the shockwave is $A(p_1 - p_2)$, and the rate of the change of momentum is:

$$\Delta(\dot{m}v) = A\rho_2 v_2 v_2 - A\rho_1 v_1 v_1 \tag{7.25}$$

So Equation (7.24) becomes:

$$A(p_1 - p_2) = A(\rho_2 v_2 v_2 - \rho_1 v_1 v_1)$$

Then:

$$p_1 + \rho_1 v_1^2 = p_2 + \rho_2 v_2^2 \tag{7.26}$$

Combining Energy equation (7.7) and (7.13) for the airflow before and after the shockwave:

$$\frac{a_1^2}{\gamma - 1} + \frac{v_1^2}{2} = \frac{a_2^2}{\gamma - 1} + \frac{v_2^2}{2} = \frac{\gamma + 1}{2(\gamma - 1)} a_c^2 \tag{7.27}$$

Rewrite Equation (7.27) with the expression of speed of sound and the Ideal Gas Law:

$$\frac{p_1}{\rho_1} = \frac{\gamma + 1}{2\gamma} a_c^2 - \frac{(\gamma - 1)v_1^2}{2\gamma} \tag{7.28}$$

and:

$$\frac{p_2}{\rho_2} = \frac{\gamma + 1}{2\gamma} a_c^2 - \frac{(\gamma - 1)v_2^2}{2\gamma} \tag{7.29}$$

Divide the left side of Equation (7.24) by the left side of the Continuity equation (7.23), $\rho_1 v_1$, and divide the right side of Equation (7.24) by the right side of the Continuity equation (7.23), $\rho_2 v_2$:

$$\frac{1}{v_1}\left(\frac{p_1}{\rho_1} + v_1^2\right) = \frac{1}{v_2}\left(\frac{p_2}{\rho_2} + v_2^2\right) \tag{7.30}$$

Substitute Equations (7.28) and (7.29) into Equation (7.30):

$$\frac{1}{v_1}\left(\frac{\gamma + 1}{2\gamma} a_c^2 - \frac{(\gamma - 1)v_1^2}{2\gamma} + v_1^2\right) = \frac{1}{v_2}\left(\frac{\gamma + 1}{2\gamma} a_c^2 - \frac{(\gamma - 1)v_2^2}{2\gamma} + v_2^2\right)$$

Simplify:

$$\frac{1}{v_1}\left(\frac{\gamma + 1}{2\gamma} a_c^2 + \frac{(-\gamma + 1 + 2\gamma)v_1^2}{2\gamma}\right) = \frac{1}{v_2}\left(\frac{\gamma + 1}{2\gamma} a_c^2 + \frac{(-\gamma + 1 + 2\gamma)v_2^2}{2\gamma}\right)$$

$$\Downarrow$$

$$\frac{1}{v_1}\left(\frac{\gamma + 1}{2\gamma} a_c^2 + \frac{(\gamma + 1)v_1^2}{2\gamma}\right) = \frac{1}{v_2}\left(\frac{\gamma + 1}{2\gamma} a_c^2 + \frac{(\gamma + 1)v_2^2}{2\gamma}\right)$$

$$\Downarrow$$

$$\frac{1}{v_1}(a_c^2 + v_1^2) = \frac{1}{v_2}(a_c^2 + v_2^2) \tag{7.31}$$

Reorganize (7.31) to get: $\frac{1}{v_1} a_c^2 - \frac{1}{v_2} a_c^2 = v_2 - v_1$

$$\Downarrow$$

$$\frac{1}{v_1 v_2} a_c^2 (v_2 - v_1) = v_2 - v_1$$

$$a_c^2 = v_1 v_2$$

That is:

$$M_1^* M_2^* = 1 \tag{7.32}$$

Equation (7.32) shows clearly that the speed coefficient after a normal shock-wave is less than "1", $M_2^* = \frac{1}{M_1^*} < 1$, if the speed coefficient before the shock-wave is greater than "1", $M_1^* > 1$. The speed coefficient is not Mach number, but it can indicate whether the airflow is subsonic or supersonic, as shown in Figure 7.3. The airflow must be supersonic before a shockwave ($M_1 > M_1^* > 1$), and Equation (7.32) states that the airflow must be subsonic after a normal shockwave ($M_2 < M_2^* < 1$), and the higher M_1 is, the lower M_2 will be, which means the higher air speed before a normal shockwave is, the lower air speed after the shockwave will be.

Using Equation (7.15) to express M_1^* and M_2^*, substitute both into Equation (7.32) to obtain the equation to calculate the Mach number M_2 after a normal shockwave with the Mach number M_1 before the shockwave:

$$M_2^2 = \frac{2 + (\gamma - 1)M_1^2}{2\gamma M_1^2 - (\gamma - 1)} \tag{7.33}$$

Use (7.32) to obtain:

$$M_2^* = \frac{v_2}{a_c} = \frac{1}{M_1^*} \Rightarrow \frac{v_2}{v_1} = \frac{a_c}{v_1 M_1^*} = \frac{1}{(M_1^*)^2}$$

Apply Equation (7.15) to this equation obtains the relationship of airspeeds before and after the normal shockwave, v_1 and v_2:

$$\frac{v_2}{v_1} = \frac{2}{(\gamma + 1)M_1^2} + \frac{\gamma - 1}{\gamma + 1} \tag{7.34}$$

Applying Equation (7.34) to the Continuity equation for a normal shockwave (7.23), we obtain the relationship of air density before and after the normal shockwave, ρ_1 and ρ_2:

$$\frac{\rho_2}{\rho_1} = \frac{(\gamma + 1)M_1^2}{2 + (\gamma - 1)M_1^2} \tag{7.35}$$

To find the ratio of air temperature before and after a normal shockwave, T_1 and T_2: set the ratio of Mach number before and after a normal shockwave, M_1 and M_2, first:

$$\frac{M_1}{M_2} = \frac{v_1 a_2}{v_2 a_1} = \frac{v_1}{v_2}\sqrt{\frac{\gamma R_M T_2}{\gamma R_M T_1}} \Rightarrow \frac{T_2}{T_1} = \frac{M_1^2}{M_2^2}\frac{v_2^2}{v_1^2}$$

Then substitute Equation (3.33) and Equation (7.34) into this equation:

$$\frac{T_2}{T_1} = 1 + \frac{2(\gamma-1)(\gamma M_1^2+1)}{(\gamma+1)^2 M_1^2}(M_1^2-1) \tag{7.36}$$

Using the Ideal Gas Law to obtain the ratio of air pressure before and after the normal shockwave, p_1 and p_2:

$$\frac{p_2}{p_1} = \frac{\rho_2 T_2}{\rho_1 T_1}$$

Substitute Equation (7.35) and Equation (7.36) into this equation:

$$\frac{p_2}{p_1} = \frac{2\gamma M_1^2 - (\gamma-1)}{\gamma+1}. \tag{7.37}$$

Alternatively, using Equations (7.30), (7.5), and (7.35) can derive Equation (7.37) as well.

According to the condition of a shockwave formation, the Mach number before a shockwave, M_1, must be greater than "1", so the ratios of $\frac{\rho_2}{\rho_1}$, $\frac{T_2}{T_1}$, and $\frac{p_2}{p_1}$ shown in (7.35), (7.36), and (7.37), respectively, are all greater than "1", which means that the air pressure, density, and temperature have all suddenly increased after a normal shockwave. The ratio of $\frac{v_2}{v_1}$ in Equation (7.34) is smaller than "1", which means that the airspeed decreases after a normal shockwave. It also shown that the higher M_1 before the shockwave is, the higher, the pressure, density, and temperature of air will be after the shockwave, and the lower the airspeed will become.

Example 7.3

The Mach number of airflow before a normal shockwave is $M=1.2$ at sea level conditions. (a) What is M_1^* before and M_2^* after the shockwave? (b) Find the air property: p_2, ρ_2, T_2, airspeed v_2, and M_2 after the shockwave.

SOLUTION

At sea level: $p_1 = 1.013 \times 10^5$ Pa, $T_1 = 288$ K, and $\rho_1 = 1.225$ kg/m³.

Calculate $M_1^*: M_1^* = \frac{v_1}{a_c}$, $v_1 = M_1 a_1$

$$v_1 = M_1 a = 1.2\sqrt{\gamma R_m T_1} = 1.2\sqrt{1.4 \times 287 \times 288} = 408.2(\text{ms}^{-1}).$$

Use Equations (7.8) and (7.9) to calculate a_c, then M_1^*.

Or to use Equation (7.15) to calculate M_1^* directly,

$$M_1^* = \sqrt{\frac{(\gamma+1)}{2}} \frac{M}{\sqrt{1+\frac{(\gamma-1)}{2}M^2}} = \sqrt{\frac{1.4+1}{2}} \frac{1.2}{\sqrt{1+\frac{1.4-1}{2}1.2^2}} = 1.16$$

From Equation (7.32): $M_2^* = \frac{1}{M_1^*} = \frac{1}{1.16} = 0.86$

Calculate the airspeed after the shockwave, v_2 using Equation (7.34):

$$\frac{v_2}{v_1} = \frac{2}{(\gamma+1)M_1^2} + \frac{\gamma-1}{\gamma+1} = \frac{2}{(1.4+1)1.2^2} + \frac{1.4-1}{1.4+1} = 0.745$$

and $v_2 = 408.2 \times 0.745 = 304.3$ (m/s)

The air density, temperature, and pressure after the shockwave are calculated, using Equations (7.35), (7.36), and (7.37), respectively:

$$\frac{\rho_2}{\rho_1} = \frac{(\gamma+1)M_1^2}{2+(\gamma-1)M_1^2} = \frac{(1.4+1)\times 1.2^2}{2+(1.4-1)\times 1.2^2} = 1.34$$

then, $\rho_2 = 1.34 \times 1.225 = 1.64\ (\text{kgm}^{-3})$.

$$\frac{p_2}{p_1} = \frac{2\gamma}{(\gamma+1)}M_1^2 - \frac{\gamma-1}{\gamma+1} = \frac{7M_1^2-1}{6} = \frac{7\times 1.2^2-1}{6} = 1.51$$

then, $p_2 = 1.51 \times 1.013 \times 10^5 = 1.53 \times 10^5$ (Pa).

$$\frac{T_2}{T_1} = 1 + \frac{2(\gamma-1)(\gamma M_1^2+1)}{(\gamma+1)^2 M_1^2}(M_1^2-1)$$

$$= 1 + \frac{2\times(1.4-1)\times(1.4\times 1.2^2+1)}{(1.4+1)^2\times 1.2^2}(1.2^2-1) = 1.13$$

then, $T_2 = 1.13 \times 288 = 324.9$ (K).

To calculate the Mach number after the shockwave by definition:

$$M_2 = \frac{v_2}{a_2} = \frac{304.3}{\sqrt{1.4\times 287\times 324.9}} = 0.84$$

or using Equation (7.33)

$$M_2^2 = \frac{2+(\gamma-1)M_1^2}{2\gamma M_1^2-(\gamma-1)} \Rightarrow M_2 = \sqrt{\frac{2+(\gamma-1)M_1^2}{2\gamma M_1^2-(\gamma-1)}}$$

$$= \sqrt{\frac{2+0.4\times 1.2^2}{2\times 1.4\times 1.2^2-0.4}} = 0.84$$

achieves the same result.

Shockwave in a Flow Path

In a converging-diverging nozzle, the airflow becomes supersonic in the divergent part of the nozzle. The air pressure at the exit of the nozzle, p_2, can be determined by Equation (7.22), and Equations (7.11) and (7.12), if the physical shape of the nozzle and the parameter of the airflow at the inlet of the nozzle are known. The process of the airflow passing though the nozzle is reversible (isentropic). However, the process might not be reversible if the air pressure p_b outside the outlet of the nozzle is not the p_2 as expected, in particular, $p_b > p_2$. p_b is the back pressure to the nozzle. It will produce a compression disturbance to the supersonic airflow if, $p_b > p_2$, and a normal shockwave will be formed.

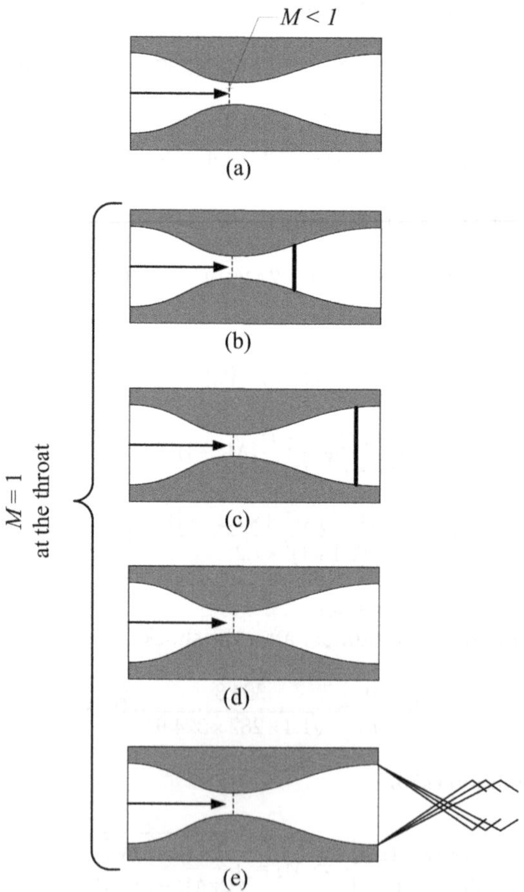

FIGURE 7.7
Airflow through a convergent-divergent nozzle with different back pressure p_b (a) v does not reach sonic at throat. v reaches speed of sound at throat in (b), (c), (d), and (e) with different back pressure p_b.

Figure 7.7 shows several scenarios of shockwaves in a convergent-divergent nozzle in relation to its back pressure p_b. (Note: p_1 is the air pressure at the inlet of the nozzle.)

(a) Airflow cannot reach a_c at the throat, and the airflow inside the nozzle is subsonic if $p_b > p_c$. There is no possibility to form a shockwave.

(b) There is a normal shockwave formed in the divergent section of the nozzle, if $p_c > p_b > p_2$. The airflow can reach the sonic state at the throat: $M = 1$, and become supersonic in the divergent section. The compression pressure disturbance originated at the exit causes a normal shockwave.

(c) Similar to (b), at the throat: $M = 1$, there is a normal shockwave formed in the divergent section of the nozzle, when $p_c > p_b > p_2$. But the position of the shockwave moves toward to the outlet of the nozzle when p_b is much lower than p_c, but still greater than p_2: $p_c >> p_b > p_2$ (p_b is close to p_2).

(d) At the throat: $M = 1$, airflow will increase its speed from subsonic at the inlet to supersonic at the outlet, and $p_c > p_b = p_2$ (ideal condition). Airspeed at the throat become sonic and there is no shockwave within the nozzle.

(e) Supersonic airflow will expand more and form expansion waves after it exits from the nozzle if $p_b < p_2$.

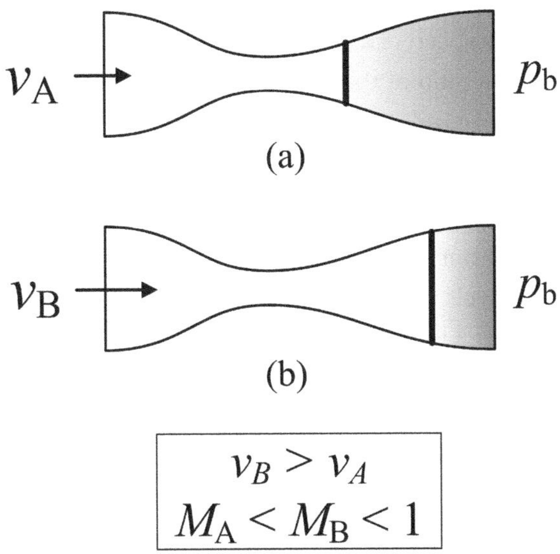

FIGURE 7.8
Airflow through a convergent-divergent nozzle with different airspeeds (a) and (b) at the inlet but the same back pressure p_b.

On the other hand, in the situation that the back pressure is constant, p_b = constant, the higher the airspeed at the inlet of the nozzle is, the further to the outlet, the shockwave moves to, as shown in Figure 7.8 (a) and (b). In both cases, the airspeed at the inlet is subsonic. According to Equation (7.37), the pressure difference across a normal shockwave increases when the Mach number before the shockwave increases. In Figure 7.8, $M_A < M_B$, and $\frac{p_b}{p_A} < \frac{p_b}{p_B}$. So a higher-speed subsonic airflow "pushes" shockwave further downstream in the nozzle.

Exercises

1. An aircraft is in flight, and its TAS = 220 m/s. The ambient temperature is $T = 253$ K. What is the stagnation temperature on its leading edge?
2. The Mach number of an airflow is $M = 0.7$. What is its speed coefficient M^*?
3. Air flows through a converging-diverging nozzle. The Mach number of airflow at the inlet of the nozzle is $M_1 = 0.75$, and the air condition is that at sea level at the inlet. The Mach number of the airflow at the outlet of the nozzle reaches $M_2 = 1.3$.
   (a) What is the speed of sound at the throat a_c?
   (b) What is the ratio of the cross-section area at the inlet to that at the throat?
   (c) What is the airspeed and air temperature at the outlet of the nozzle?
4. The Mach number before a normal shockwave is $M = 1.4$. What is the Mach number after the normal shockwave?
5. An airflow travels at $M = 1.5$ at sea level conditions before a normal shockwave. Find out the air property and airspeed after the shockwave.

8

Aerodynamics of Transonic Aerofoils

For a transonic flight, the free-stream Mach number of an aircraft reaches or exceeds its critical Mach number when local airflow becomes sonic, or supersonic, normally at the most cambered location over its aerofoil. As we learned in Chapter 7, a shockwave can be formed when the airflow is supersonic with compression pressure disturbance. The air pressure will be increased suddenly after a shockwave. This sudden change in air pressure will affect the lift of on the transonic wing and the aircraft's performance will be affected.

To understand the features of airflow around transonic aircraft, we will analyze the development of shockwave over aerofoil; the effects of shockwave on lift, drag, and stall; and the effects on flight control.

Shockwaves on Aerofoil

When the free-stream Mach number M_{fs} reaches the critical Mach number, M_{crit}, the airflow at a location over aerofoil is sonic, and Mach waves occur at that location due to the variation in local air pressure. The speed of local air is greater than the speed of sound, when the free-stream Mach number increases or the AoA increases. The airflow near the aerofoil is similar to that in a convergent-divergent nozzle (a half nozzle) as shown in Figure 8.1. The line T in Figure 8.1 is the location of the "throat", the critical point, where $M=1$. The speed of the air after T would be greater than the local speed of sound, $M_L>1$, and the local air pressure p_L decreases, and $p_L<p_{fs}$. The air pressure behind the trailing edge of the aerofoil should be p_{fs}, so a normal shockwave wave will be formed on the upper surface of the aerofoil (as was discussed in Chapter 7).

Structure of Shockwave on Aerofoil

Figure 8.2 shows a normal shockwave at the upper surface of an aerofoil: the airflow is supersonic just before the shockwave; however, free-stream airspeed is lower than the speed of sound; the airflow behind the shockwave becomes subsonic, and air pressure, density, and temperature increase after the shockwave.

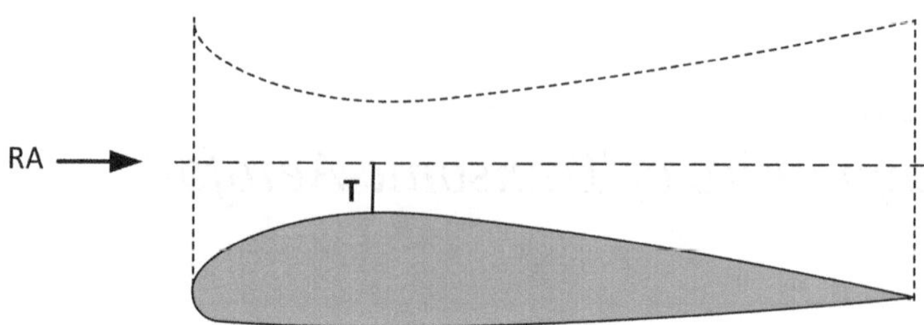

FIGURE 8.1
Compressible airflow over aerofoil comparing with convergent-divergent nozzle.

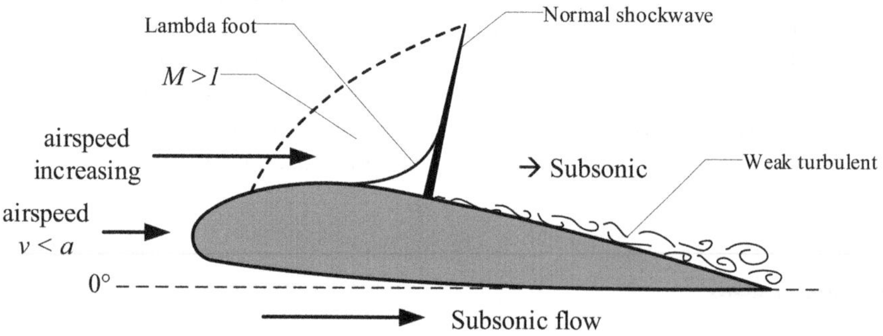

FIGURE 8.2
Feature of the airflow structure before and after a normal shockwave on aerofoil.

The schematic normal shockwave in Figure 8.2 indicates that the root part of the shockwave is not a simple plain thin wave front, but it is a wider structure. This part of shockwave is called *lambda (λ) foot*. One side of "λ" is the part of the normal shockwave, and the other side of "λ", in front of the normal shockwave, is a small oblique shockwave. The lambda foot is partially caused by "pressure leak" – pressure change can propagate through the boundary layer, where the local air particle speed can be lower than speed of sound. The pressure change in boundary layer is with a gradient, is not a sudden change, so the pressure in the "λ" is greater than the pressure in front of the oblique part of the shockwave, but smaller than the pressure behind the normal shockwave.

There is a turbulent wake behind a normal shockwave on the aerofoil, as shown in Figure 8.2. The turbulent wake oscillates. The higher the Mach number is before the normal shockwave, the stronger the shockwave is, and then the stronger the turbulent wake's oscillation will be. The turbulent wake tends to separate from the surface. The separation can occur at the rear part of the aerofoil if the shockwave is relatively weak. When the shockwave gets stronger, the separation point moves forward – it gets closer to the shockwave.

Movement of Shockwave on Aerofoil

Normal shockwaves move along the surfaces of an aerofoil, when the free-stream Mach number changes.

Figure 8.3 shows the general movement of shockwaves when the free-stream Mach number, M_{fs}, increases from, M_{crit}:

When M_{fs} reaches the critical Mach number, M_{crit}, the local airflow becomes sonic, *i.e.* the local Mach number $M_L = 1$, and Mach waves occur at that location Figure 8.3 (a).

When M_{fs} increases from M_{crit}, a normal shockwave forms on the upper surface of the aerofoil as shown in Figure 8.3 (b). The dashed line marks the region, where the airflow is supersonic, $M_L > 1$.

As M_{fs} increases further, the normal shockwave moves rearwards along the upper surface of the aerofoil, as shown in Figure 8.3 (c). The intensity of the shockwave increases, and the lambda foot of the shockwave is getting greater than that in (b). The turbulent wake behind the shockwave oscillates more, and the separation point moves forward.

When M_{fs} continues to increase, a normal shockwave forms on the lower surface of the aerofoil, as shown in Figure 8.4 (d). The upper shockwave gets stronger, and its turbulent wave separates just behind the shockwave.

M_{fs} increases more from the situation (d), and both the upper and lower shockwaves move rearwards, and the oscillation of the turbulent wakes behind both shockwaves become stronger. However, the normal shockwave on the lower surface moves faster than the upper one does as shown in Figure 8.3 (e), and will reach the trailing edge of the aerofoil first.

When M_{fs} increases further from (e), close to $M_{fs} \approx 1$, the upper shockwave moves to the trailing edge, and then both shockwaves settle at the trailing edge, shown in Figure 8.3 (f). Now, both upper surface and lower surface are covered by supersonic airflow, $M > 1$, and there is no separation on either surface of the aerofoil.

When M_{fs} increases again from (f), and becomes greater than 1, a bow shockwave forms at the leading edge, and detaches from the leading edge, as shown in Figure 8.3 (g). The free-stream Mach number M_{fs}, when the bow shockwave forms and detaches from the leading edge, is called *Detachment Mach Number* – M_{det}.

The bow shockwave consists of one piece of normal shockwave (in the middle) and two pieces of oblique shockwave (one at each side). In the situation shown in Figure 8.3 (g): $M_{\text{fs}} > 1$, the Mach number of the airflow behind the bow shockwave and in front the leading edge is subsonic (behind the normal part of the bow shockwave). The airflow behind of the oblique parts of the bow shockwave is still supersonic. Meanwhile, supersonic flow is over both surfaces of the aerofoil.

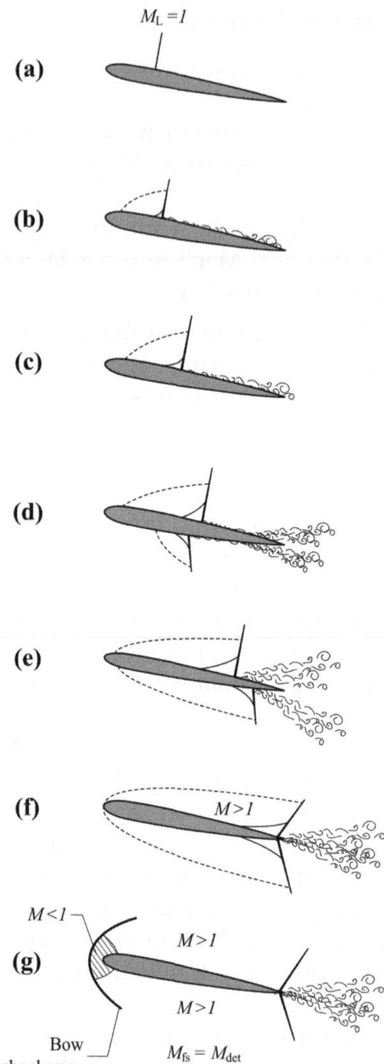

FIGURE 8.3
The development of shockwaves around aerofoil when M_{fs} increases from M_{crit} to M_{det}, from (a) to (g).

For a symmetrical aerofoil sitting at a 0° AoA, shockwave formation and movement on upper and lower surfaces occurs simultaneously when the free-stream Mach number change from M_{crit} to M_{det}.

The aerodynamic feature described here is the airflow feature over a transonic aerofoil. The Mach number range for transonic flights is from M_{crit} to M_{det}.

Effective Critical Mach Number

Each aircraft should have its designed or nominated critical Mach number, $M_{crit,}$ specified by its manufacturer. An aircraft can undertake different maneuvers apart from level flight at an airspeed close to M_{crit}, when increase of AoA of an aerofoil is required. This increase of AoA can produce a local airspeed, which make its $M_L = 1$, over the aerofoil. Some of other maneuvers, for example, rolling, banking, or descending, might increase the local airspeed over some parts of the aircraft body to make the $M_L = 1$. Once $M_L = 1$, the M_{fs} becomes the effective M_{crit}. In those cases, the local Mach number reaches "1" before the free-stream Mach number of the aircraft reaches its designed M_{crit}, so the effective M_{crit} is smaller than the designed/nominated M_{crit}. In other words, increase of AoA and maneuverers might decrease M_{crit}.

Changes of CP, C_L, and C_D on a Transonic Aerofoil

The air property around the aerofoil changes significantly when shockwaves are formed on aerofoil. As was discussed in Chapter 7, there is a sudden increase in air pressure, density, and temperature when airflow passes through a normal shockwave. The turbulent wake behind the shockwave oscillates and separates from the surface of aerofoil – boundary separation, when the free-stream Mach number M_{fs} increases. All of these facts affect the aerodynamic forces produced by the aerofoil.

Shockwave on Lift

The formation and movements of shockwaves on an aerofoil cause significant changes of the pressure distribution around the aerofoil. Figure 8.4 shows

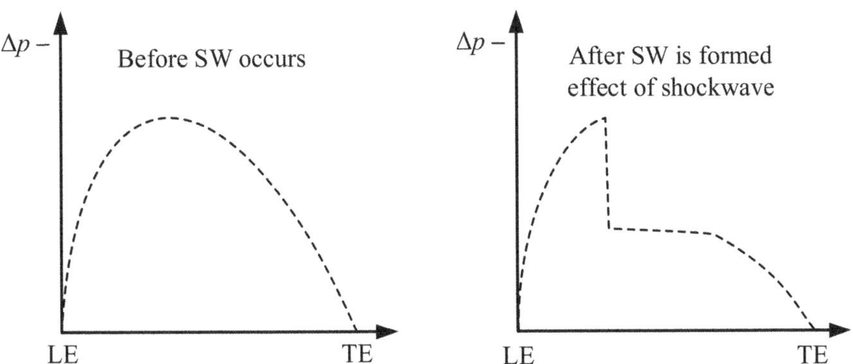

FIGURE 8.4
(a) Pressure distribution over aerofoil before a shockwave occurs; (b) Pressure distribution over aerofoil after a shockwave is formed.

the distributions of relative pressure over the upper surface of an aerofoil: (a) is that without a shockwave (SW), and (b) is that with a shockwave (SW). There is a sudden increase in pressure in Figure 8.4 (b).

The ability to produce lift is weakened after a normal shockwave formed over the aerofoil, so the lift coefficient, C_L, of the aerofoil decreases. The level of decrease in C_L depends on the intensity of the shockwave. The stronger the shockwave is, the higher the pressure after the shockwave is, and the more the C_L decreases.

The center of pressure, CP, will change since the pressure distribution has changed after the shockwave. CP will move forward initially because the shockwave starts at the location closer to the LE. CP will move rearward if the shockwave moves rearward.

Shock Drag

The formation of shockwaves and the separation of the turbulent wakes behind shockwaves on an aerofoil contribute a sudden increase to the total drag. The additional drag caused by shockwaves is called *Shock Drag*. There are two types of shock drag: *wave drag* and *boundary separation drag.*

Wave Drag

When a shockwave is formed, the air pressure, temperature, and density are all increased suddenly. Energy is required to form and maintain the shockwave, and the energy is provided by the airflow. The higher the Mach number before the shockwave is, the more kinetic energy the airflow would lose to overcome the "resistance" to pass through the shockwave. This "resistance" is the *wave drag,* also called energy drag.

Boundary Separation Drag

The turbulent wake detaches from the surface of an aerofoil behind the shockwave, when the free-stream Mach number M_{fs} increases. This separation affects the aerodynamics forces just like in subsonic boundary layer separation – causing a type of form drag, which is the *boundary separation drag.*

Figure 8.5 illustrates the extra portions in drag caused by shockwaves and their change with the free-stream Mach number. The lower dashed line in Figure 8.4 shows the total drag without shockwaves; the drag rises to the solid line due to the formation of shockwave on the aerofoil when $M_{fs} > M_{crit}$. The gap between the solid line and the dashed line is wave drag. The shaded area over the solid line represents the boundary separation drag. It is clear that the drag level increases significantly once the turbulent wake separates from the aerofoil surface.

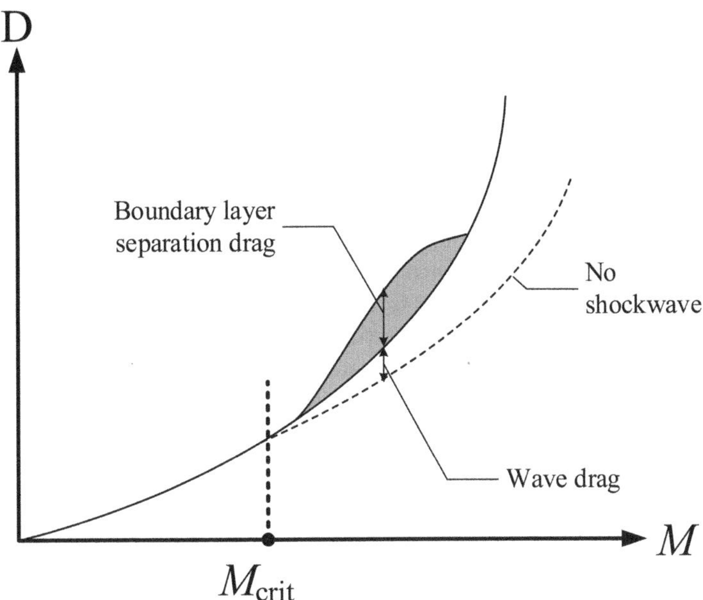

FIGURE 8.5
Shock drag consisting of wave drag and boundary layer separation drag.

Changes of CP, C_L, and C_D between M_{crit} to M_{det}

The intensity of normal shockwaves increases and their locations on aerofoil change along the aerofoil when the M_{fs} of the aircraft increases from M_{crit} to M_{det}. Of course, CP, C_L, and C_D will change with the change of the shockwaves as well.

We will follow the movements of the shockwaves on an aerofoil illustrated in Figure 8.3 and analyze the changes of center of pressure, CP, the lift coefficient, C_L, and drag coefficient, C_D, step by step.

In Figure 8.3, free-stream Mach number M_{fs} increases from M_{crit} (a) to M_{det} (g).

From (a) to (b): the local airspeed is just greater than local speed of sound, the local pressure is relatively low, and a shockwave is forming over the aerofoil. The CP moves rearward a little and C_L increases due to the decrease of the local pressure. But the C_D increases slightly because a normal shock is forming and it is weak at this stage.

From (b) to (c): The free-stream Mach number M_{fs} increases, and the normal shockwave on the upper surface of the aerofoil becomes little stronger and starts moving rearward slowly. Now there is a sudden increase in air pressure behind the shockwave, and the turbulent wake behind the shockwave becomes active, and at the end of this period (c), it start to separate near the trailing edge. CP is moving forward more, and the C_L starts decreasing because of the sudden increase of air pressure on the upper surface of the aerofoil. Meanwhile, C_D increases clearly because the stronger shockwave means more kinetic energy is dissipated in the formation of the shockwave, more wave drag.

From (c) to (d): M_{fs} increases further during this period. The normal shockwave on the upper surface is getting stronger, and there is a higher sudden increase of air pressure. Another normal shockwave starts forming with its turbulent wake on the lower surface of the aerofoil. So C_L decreases in this period rapidly. The CP first moves forward, and is at the forward-most position when the normal shockwave initially appears on the lower surface. Then CP starts moving rearward when both shockwaves move toward to the trailing edge with the increase of M_{fs}. The turbulent wake behind the shockwave oscillates and the separating point of the turbulent wake starts moving closer to the shockwave. The rate of C_D increase during this period is much higher because the upper shockwave and its turbulent wake is getting stronger and the lower shockwave is formed.

M_{fs} increases to the stage shown in Figure 8.3 (d), C_D increases significantly (the rate of C_D increase is at its maximum). This free-stream Mach number, M_{fs}, is called *critical drag rise Mach number* M_{cdr} (or *drag-divergence Mach number* M_{dd}).

From (d) to (e): During this M_{fs} increase period, both shockwaves are getting stronger and move rearwards. But the lower shockwave is moving faster than the upper one. CP moves rearward and C_L starts a gradual increase from the sharp fall, because the supersonic region in front of the shockwaves, where pressure distribution establishes and produces aerodynamic forces, is extended toward to the trailing edge. However, C_D continues to increase, because the intensity of the two shockwaves increase with the increase of M_{fs}, and their turbulent wakes are separating more actively behind the shockwaves. There are higher wave drags and stronger boundary layer separation drags in this period and C_D may reach its peak.

From (e) to (f): The lower shockwave moves further and settles at the lower trailing edge of the aerofoil first. Then the upper shockwave follows and settles at the upper trailing edge with the increases of M_{fs}. In this period, CP continues moving rearward, and settles at the middle of the aerofoil, approximate 50% of the chord, when both shockwaves have located at the trailing edge. Meanwhile, C_L increases because the proportion of the aerofoil surface behind shockwaves, where there is a sudden pressure increase, is getting smaller in this period. The airspeed in front of the shockwaves is supersonic and air pressure decreases, in particular, in front of the shockwave of upper surface.

C_D decreases because the separating points of the turbulent wakes move to the rear end, and eventually there is no separation on both sides of the aerofoil in this period. So the boundary layer separation drag in the shock drag decreases, and the wake drag in the shock drag still remains.

From (f) to (g): M_{fs} increases to M_{det}, as in Figure 8.6 (g), and a bow shockwave forms and detaches from the leading edge of the aerofoil. Both of the upper and lower normal shockwaves stay at the trailing edge in this period. Both surfaces are gradually covered with supersonic airflow, and the air

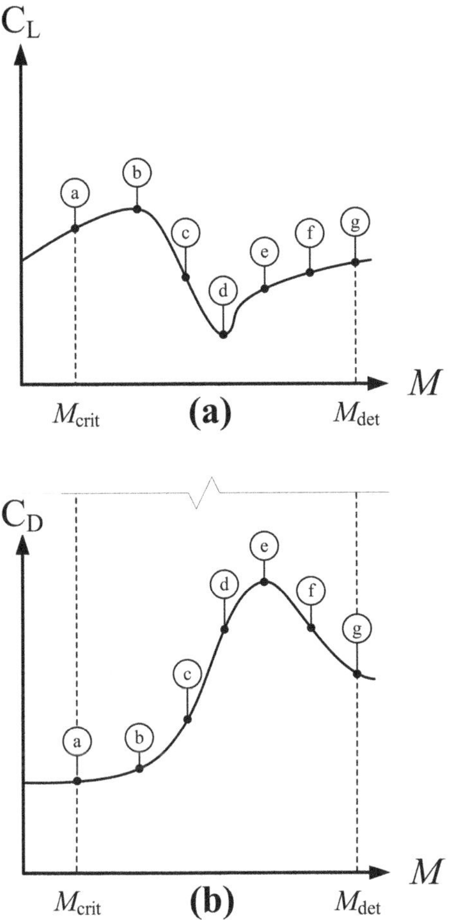

FIGURE 8.6
(a) Change of the lift coefficient, C_L, with the increase of M_{fs} from M_{crit} to M_{det}; (b) Change of the drag coefficient, C_D, with the increase of M_{fs} from M_{crit} to M_{det}.

pressure on both surfaces could be lower than that in free-stream, p_{fs}. The pressure on upper surface is little lower than the pressure on the lower surface, and the profiles of pressure distribution over the upper surface and under the lower surface are very similar. So the CP remains at the position of approximate 50% of chord, and C_L decreases slowly with the increase of M_{fs} due to the formation of the bow shockwave and the similarity of the pressure distribution. C_D decreases a little in this period, as the position of the normal shockwaves has not changed, and there is no turbulent wake separation at all on the aerofoil during this period.

The summary of the changes of C_L and C_D with the increase of M_{fs} from M_{crit} to M_{det} is shown in Figure 8.6 (a) and (b), which is correlated to the shockwave movement shown in Figure 8.3. For each design of aerofoil with a specified camber and thickness, the magnitude of the changes of its CP, C_L, and C_D may vary. However, in general, the trends of the changes are:

- CP moves rearward when the both shockwaves move toward the trailing edge of the aerofoil and settles near the central part of the aerofoil;
- The lift coefficient C_L will decrease after the initial increasing soon after M_{fs} reaches M_{crit}, because the increase of air pressure after the upper normal shockwave. The stronger the shockwave become, the more forward the turbulent wake moves, and the more the C_L decreases;
- The drag coefficient C_D will increase when shockwaves are formed, and this changes significantly when the turbulent wake separation takes place;
- Both C_L and C_D can recover from their low and high values, respectively, when shockwaves reach the area near the trailing edge.

Note that the change of drag, D, could be different from that of drag coefficient, C_D, as shown in Figure 8.6 (b). When M_{fs} is close to M_{det}, the drag is still increases obviously, even though the drag coefficient is slightly decreasing. Remember drag is proportionally to v^2, and at $M_{fs} \approx M_{det}$, the airspeed is very significant.

Shock Stall

As discussed previously, the lift coefficient C_L and drag coefficient C_D changes with the development and movement of normal shockwaves formed on the upper and lower surfaces of a transonic aerofoil, when the free-stream Mach number M_{fs} of the airplane increases. The intensity of the shockwaves gets stronger, and the turbulent wakes behind the shockwaves separate from the surface of aerofoil with the separating point moving close to the shockwaves when M_{fs} increases.

A turbulent wake produces vortices, which detach from the wake alternatively with a certain frequency. The detachment of the vortices causes noise, oscillation of the shockwave, and the increase in C_D. When the M_{fs} increases to a curtain level, the detachment or separation becomes very intense, and happens immediate after the shockwave, the aircraft experiences the random vibration with loud noise, which is called *Mach buffet*, or *sonic buffet*.

The pilot's experience is like that of a low-speed/high-AoA buffeting when the aircraft is in Mach buffet.

The ability to produce lift is weakened, when a shockwave forms and the turbulent wake separates over the aerofoil. The aerofoil will fail to produce lift, if the intensity of the shockwaves keeps on increasing after Mach buffeting, and the aircraft will be in stall. That is *shock stall.*

The lift coefficient of the aerofoil will be at its lowest, when the aircraft is near its shock stall. At the meantime, its drag coefficient is likely increasing sharply. The vibration caused by intensive Mach buffet could lead to the failure of control of the flight. So Mach buffet should be avoided, and there should be audial and visual warning for the buffet.

Shockwaves on Control Surfaces

The intention of deflecting a control surface is to produce a force on the control surface in order to perform curtain maneuvers. The local airflow speed can increase when the surface deflects. So normal shockwaves can form on control surfaces of a transonic aircraft if the control surface deflects. The development and movement of the shockwave would be in the same way as that on an aerofoil with the increase of M_{fs}, as discussed earlier.

The shockwave would form at the hinge area of a control surface first, for example, on the hinge of elevator, or on the hinge of rudder, and then the shockwave will move rearward on the control surface with an increase of M_{fs}. The pressure will be increased behind the shockwave, which makes the control feel heavy. The aerodynamic force exerted on the control surface due to the shockwave could be too great for the pilot to move the control surface easily. The change of pressure distribution over the control surface due to the formation of a shockwave causes ineffectiveness of the control surface, or even produces the opposite effect from the initial control intention.

There is a turbulent wake behind a normal shockwave on a control surface, and the turbulent wake separation causes the vibration of the control surface, and, in turn, the vibration of the control surface can introduce disturbances, which, in turn, leads to unstable behavior.

Transonic Control Issues

Longitudinal

An aircraft in a transonic flight experiences nose-down pitch – i.e. the nose becomes heavy due to the shockwave formed over its wing. When the nose becomes heavy, this is called *Mach tuck.* When M_{fs} increases, and the shockwaves over the wings of a transonic aircraft moves rearward, the center of pressure CP over the wing moves rearward with the shockwaves. The move of CP increases the distance between CP and CG, as shown in Figure 8.7. So, the nose-down pitch moment increases.

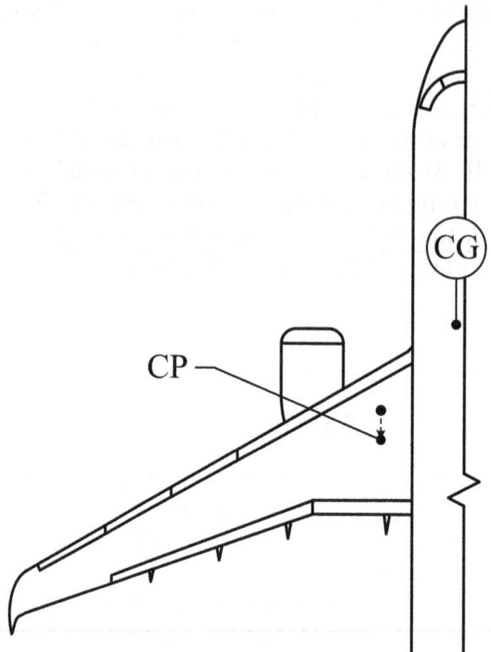

FIGURE 8.7
The rearward movement of CP when the aircraft is at transonic speed.

When the turbulent wake behind the shockwave separates earlier with the increase of M_{fs}, the down load from the wing onto the tail plane is reduced if the tail plane is not T-tail. The tail plane will produce less recovering nose-up pitch moment to overcome the pitch-down moment caused by the movement of the shockwave on the wing. So the transonic aircraft at high airspeed can experience nose heavy. The nose-down motion causes the airplane to accelerate. This acceleration leads to M_{fs} increase further, and worsens the control difficulties.

When a shockwave forms on an elevator:

- It feels heavy to move the elevator, because the shockwave sets on the elevator with the sudden increase of pressure on the surface of the elevator;
- The airplane does not response to the elevator's movement effectively, because the pressure distribution has changed over the elevator now, and it cannot produce efficient aerodynamic force for which it was designed;
- The elevator vibrates and "buzzes" noisily, or buffets, due to the turbulent wake separation behind the shockwave;
- *Adverse "stick force"*: the push force on the control column at higher M_{fs} in transonic region results in "pull" action instead of "push"

action. This is caused by the shockwave formed on the elevator. When the control column is pushed, the elevator deflected in order to produce lift – nose-down pitch, but pressure behind the shockwave on the elevator increases significantly, in particular, if there is turbulent separation, so the "lift" produced by the elevator now is a negative lift – nose-up pitch – "pull". It is risky, and confusing, when the adverse "stick force" occurs.

There are a number of design options for transonic elevators can be used to deal with the control difficulties:

- Use a thin tail plane, or relatively sharp leading-edge design to increase M_{crit} of elevator, delaying the formation of a shockwave on the elevator;
- Use Mach trim system. The input of Mach trim provides control with the aerodynamic force in the other direction to correct the adverse "stick force", and eliminates the confusion, and also can reduce Mach tuck;
- All movable slabs of elevator seen in Figure 8.8 can be used. Each slab can be operated individually. So the slabs can provide pilots with more choices of actions in longitudinal control to reduce the effect of shockwaves on the control;
- An adjustable and power-operated tail plane can be used to overcome the extra forces exerted on the elevator due to the movement of shockwave on the elevator so that it can continue to respond to the demand from the control column.

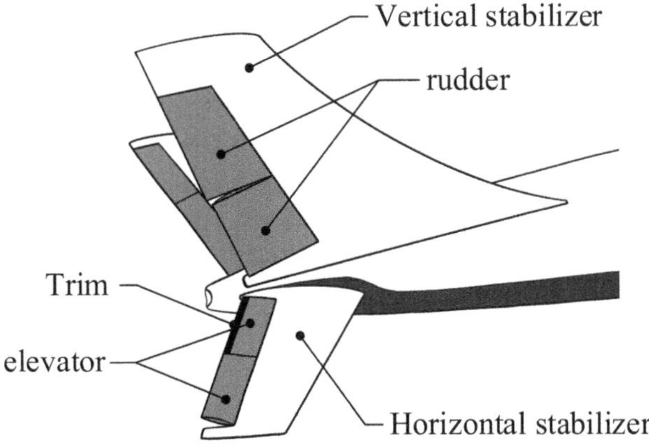

FIGURE 8.8
The controls on a tail plane of a transonic aircraft.

Lateral

When an aircraft travels in a transonic airspeed, and shockwaves can form on the surface of ailerons:

- the ailerons cannot operate effectively, since the pressure increases behind the shockwave;
- ailerons flutters and vibrates when the turbulent wake separates behind the shockwave;
- the aircraft will be in roll disturbance, because the fast vibration of ailerons varies the lift on the aerofoil randomly;
- the sudden increase in pressure and the boundary layer separation (by turbulent wake) after a normal shockwave on aileron can result in a "lift" in the opposite direction as it is intended to produce, when the aileron is deflected to perform banking. Then, the aircraft can start banking to the opposite direction, which is *aileron reversal.* (Aileron reversal can also occur when a greater "lift" is produced by deflecting aileron. The "lift" twists the wing about its lateral axis, if the wing is not rigid enough. This twist would change the angle of attack in the opposite direction. So the wing that should produce more lift to perform "bank to the right" would reduce its lift, which would make the aircraft "bank to the left".)

There are a few design options associated with a wing and its ailerons that can be used to deal with the control difficulties:

- Vortex generators are installed upstream of ailerons to delay the formation of shockwaves, and reenergize the airflow over the control surface to delay the boundary layer (turbulent wake) separation;
- Small outboard ailerons, shown in Figure 8.9. With outboard ailerons, an aircraft requires less "lift" to produce roll moment to perform banking. That will reduce the deflecting angle, and it will operate in a less shockwave and turbulent wake-affected area;
- Inboard (mid-chord/mid-span) spoilers shown in Figure 8.9, which disrupt air flow to reduce the lift on the roll-downward wing (instead of an aileron to increase lift on the roll-upward wing) to avoid aileron reversal.

Directional

When a shockwave is first formed on the hinge of rudder in a transonic flight, the function in directional control will be affected. The directional control would experience the similar difficulties as the control in other directions:

- Ineffective rudder function caused by the shockwave. The rudder is behind shockwave, so the it cannot affect the aerodynamic forces on the tail/fin to produce yaw moments;
- It feels heavy to move the rudder, as the shockwave moves rearward on the rudder, when M_{fs} increase;
- The turbulent wake and the increase of pressure behind the shockwave on rudder cause oscillation. That produces yawing disturbance. With the weakened directional control, this yawing disturbance could lead to Dutch roll.

There are a few methods to use to assist the directional stability control in a transonic flight:

- To install yaw dampers to reduce directional oscillation and to weaken/eliminate Dutch roll;
- Conventional fin and rudder combination with powered control surface to increase the effectiveness of rudder;
- All movable slabs of fin/rudder on the vertical stabilizer, as seen in Figure 8.8, can be operated separately to avoid the effects of the shockwave on rudder.

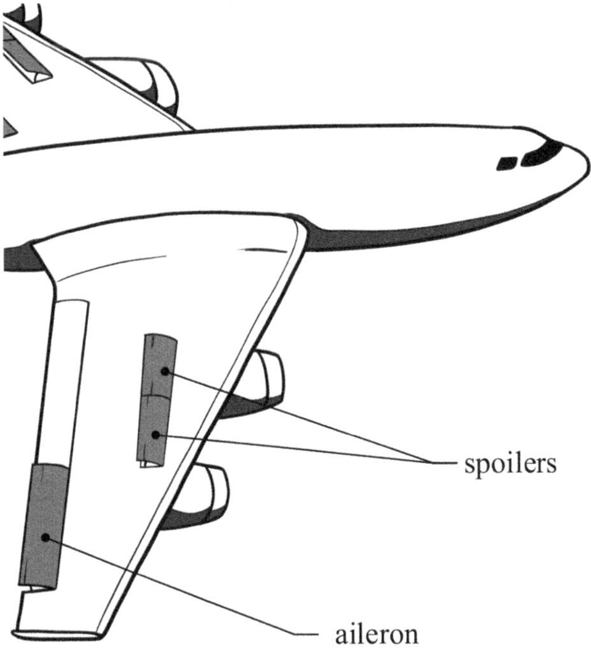

FIGURE 8.9
Lateral controls on a transonic wing.

Exercises

1. The following table gives the values of the drag coefficient of a thin aerofoil at different free-stream Mach numbers. Calculate the drag of the aerofoil at sea level condition with these Mach numbers. The area of the aerofoil is 40 m². Plot a graph of drag D and drag coefficient C_D vs. Mach number M_{fs}.

M_{fs}	0.6	0.7	0.75	0.8	0.85	0.9	0.95	1.0	1.05	1.1	1.2	1.3	1.4
C_D	0.011	0.011	0.016	0.026	0.05	0.068	0.063	0.055	0.046	0.04	0.034	0.03	0.028
Drag													

(The values in the table are modified from the example in Kermode (2012)

2. How do shockwaves move over an aerofoil as the free-stream Mach number increases from M_{crit} to M_{det}?
3. Explain why CP moves over an aerofoil when the free-stream Mach number increases from M_{crit} to M_{det}.
4. What causes the significant rise in the drag coefficient as the free-stream Mach number increases?
5. What is a bow shockwave? Explain the difference between the Mach number at the "Nose" of the bow shockwave and the free-stream Mach number.
6. What is Mach tuck? What is the adverse "stick force"?
7. What is the Mach buffet?

9

Transonic Flight and Aerofoils

Modern jet airliners can travel much higher and faster than ever before. They are transonic flights. A transonic flight faces issues caused by shockwaves, and aerodynamic limitations. The features of transonic aircraft have been designed to overcome the extra drag caused by shockwaves, control difficulties, and to improve aerodynamic limitations in transonic flights.

Transonic Speed Limits

One aspect of the limitations in transonic flights is the limits of airspeed, at which the aircraft can no longer produce the sufficient lift effectively to maintain flight.

There are two limits: high and low airspeed limit.

- The low speed limit of a transonic flight is the airspeed, at which the airplane is at low speed–high angle attack stall, caused by the boundary layer separation over the aerofoil, and the air pressure distribution, which produces lift over the aerofoil is interrupted and destroyed.
- The high-speed limit is the airspeed, at which the airplane starts Mach stall, caused by strong normal shockwaves and the separation of turbulent wake behind the shockwaves on the aerofoil surface. The formation of a shockwave over an aerofoil increases the local pressure suddenly, and the boundary layer separation of the turbulent wake causes the complete loss in lift and the significant increase in drag. Therefore, the level flight cannot be maintained in air.

Figure 9.1 is an example of a diagram of a speed limit (level flight) envelope. The left line is marked as "high AoA stalling speed" (approximately, IAS [indicated airspeed] = 100 kt), and it is the low speed limit. At a low airspeed, the lift coefficient has to be high to maintain the lift needed, which means a high angle of attack (AoA) is required, as discussed in Chapter 4. The airspeed displayed in Figure 9.1 is the true airspeed (TAS). The value of TAS on the low speed limit increases with the increase of altitude if the same

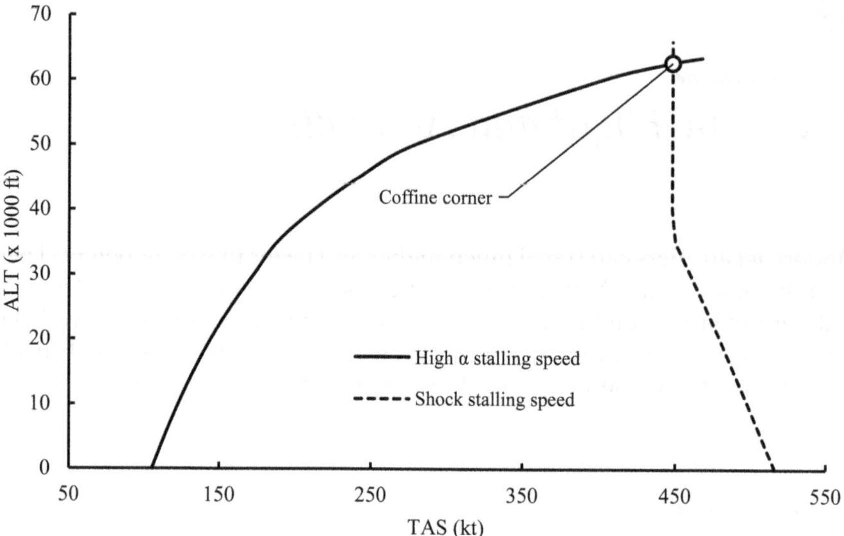

FIGURE 9.1
Speed limits.

level of lift is maintained, Lift $= C_L \frac{1}{2} \rho v^2 S$, as air density decreases with the increase of altitude. That means that the "dynamic pressure" is constant, i.e. $\frac{1}{2} \rho v^2 = \text{constant}$, at any altitude, so the indicated airspeed (IAS) displayed in a cockpit should be constant for the low speed limit at any altitude, while TAS increases.

The high-speed limit in Figure 9.1 is marked as "Shock stalling speed", which is a line of a constant Mach number (here $M = 0.78$). The character of a shockwave, including its formation, intensity, and separation, depends on the Mach number of the airflow. So, the high limit in Figure 9.1 indicates a specific aerodynamic event – Shock stall by turbulent wake separation following a constant Mach number, which is different for different aircraft.

When the Mach number is constant, the airspeed increases/decreases with the speed of sound, which changes with air temperature only. The high-speed limit is an inclined straight line under the tropopause, and is a vertical line within the tropopause, where the air temperature is constant. This line shows the change of the high-speed limit, which reflects the change of air temperature with altitude.

Coffin Corner

Figure 9.1 shows that the TAS of the low limit increases with altitude, while TAS of high-speed limit is constant when the altitude is beyond troposphere. These two limits get closer and closer while altitude increases, and then intercept at an altitude. The altitude, where the two limits intercept, is called *coffin corner.* When an aircraft reaches its coffin corner, it can only fly at one airspeed. It will be in stall if its airspeed decreases; and it will be in shock stall, if its air speed increases. An aircraft at the coffin corner is in a dangerous situation.

For the example shown in Figure 9.1, the coffin corner of that aircraft is approximately at 19,000 m (about 62,000 ft). This aircraft cannot fly at all over 19,000 m.

Buffet Boundary

When a boundary layer starts leaving the surface of an aerofoil, a street of vortices detaches from the airflow stream at a certain frequency, which causes vibration or oscillation. Therefore, an aircraft experiences in buffeting, before it is in stall. Therefore, it makes sense to mark the airspeed at the start of buffeting as *buffet boundary* to warn the pilot to avoid the aircraft from the danger of stall. Buffet boundary runs almost in parallel with the airspeed limits, as shown as the solid line of a schematic diagram in Figure 9.2. The stall airspeed limits is the dashed line. The airspeed used in Figure 9.2 is IAS (indicated airspeed), which is used in flight manual of an aircraft. The dashed lines marked as V_{MO} in Figure 9.2 is the maximum operating velocity of the aircraft.

At a lower altitude, the V_{MO} is smaller than the shock stall speed.

The speed range between the lower speed buffet boundary and the high-speed buffet boundary at each altitude is called *buffet margin, or safe*

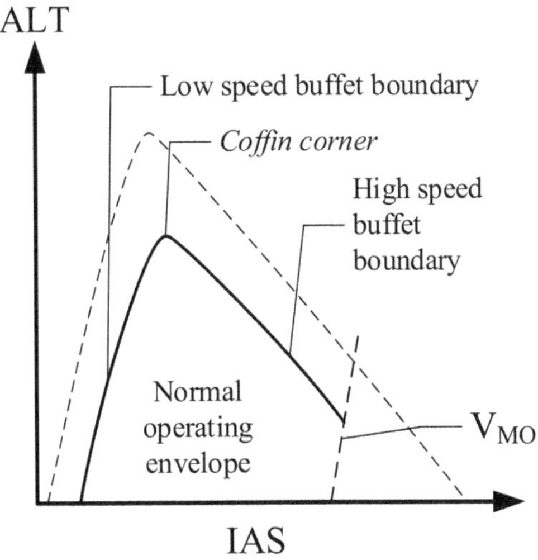

FIGURE 9.2
Schematic diagram of normal operating envelope.

speed range. The margin changes with altitude. The margins at different altitudes form a *normal operating envelope*.

The coffin corner in Figure 9.2 is the altitude, where the margin is reduced to "0." This coffin corner defined by the buffet boundaries is survivable, as the aircraft has not stalled.

Effects on Buffet Boundary

Figure 9.2 shows that the buffet boundary and buffet margin can be affected by altitude. Furthermore, the buffet boundary and margin are affected by the load an aircraft carries, and the maneuvers it performs as well. So the weight of an airplane and the load factor can change the buffet boundary and the margin.

Figure 9.3 and Figure 9.4 show the effects of the gross weight of aircraft and the load factor on buffet boundaries respectively. In Figures 9.3 and 9.4, BM stands for boundary margin, and MB80 is the boundary margin at the altitude for aircraft mass = 80,000 kg. BM70 and BM60 are the boundary margins for 70,000 kg and 60,000 kg, respectively.

When altitude increases, the buffet boundaries will get closer to each other, as shown in Figure 9.2. The buffet margin gets smaller. If the altitude increases to a certain level, the margin will be reduced to "0" – "coffin corner".

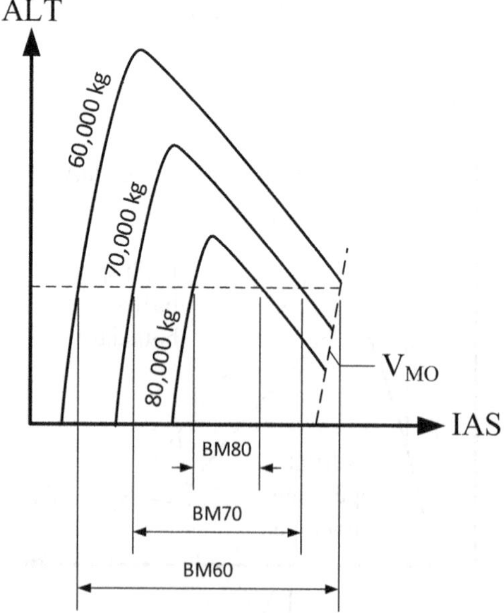

FIGURE 9.3
Change of the normal operating envelope with the gross weight.

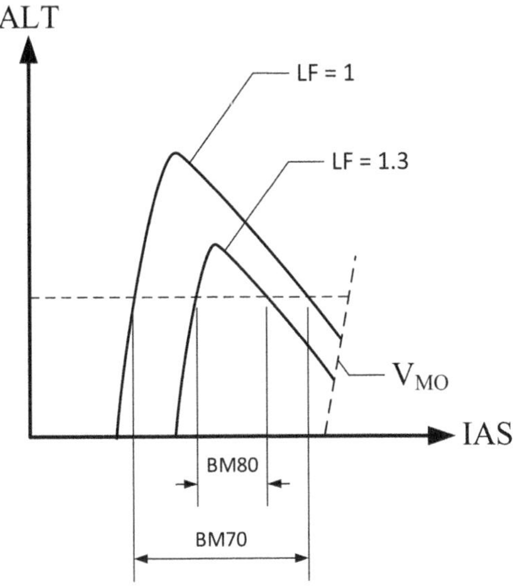

FIGURE 9.4
Change of the normal operating envelope with load factor, LF.

The boundary margin can be decreased by the increase of the weight of an aircraft as shown in Figure 9.3. It is clear that the buffet margin decreases with the increase of gross weight of aircraft. The speed on the low buffet boundary increases, but the speed on the high buffet boundary deceases, when the gross weight of aircraft increases. Figure 9.3 also shows that the "coffin corner" decreases when the weight increases, and the airspeed at the "coffin corner" increases as well.

Figure 9.4 shows that the buffet margin decreases if the load factor (LF) of an aircraft increases for the same gross weight at the same altitude. The comparison between load factor = 1 (level flight) and load factor > 1 (e.g. banking; LF = 1.3 is equivalent to banking at an angle of bank of 40°) shows that the airspeed on the low buffet boundary increases, but the speed on the high buffet boundary decreases, when the load factor becomes greater than "1"; and the "coffin corner" decreases, and the airspeed at the "coffin corner" increases, when the load factor increases.

Cross-Over Altitude

Cross-over altitude (ALT) is the altitude at which a specified IAS or CAS (IAS and CAS are very similar if there is no crosswind), and the Mach value represent the same TAS value (displayed in kts in cockpit). Above this altitude, the Mach number is used as a reference of airspeed.

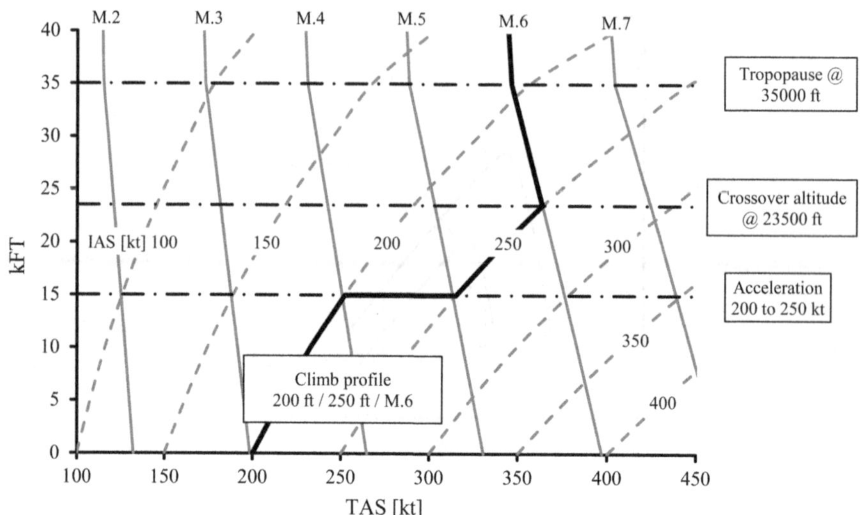

FIGURE 9.5
Cross-over altitude.

Depending on climb ALT, airspeed target, cruise speed and aircraft type, a climb profile can be planed for a flight, and the cross-over altitude is determined by the climb profile. The cross-over altitude can be different with a different climb profile for the same aircraft.

To find the cross-over altitude, for example, following the climb profile of an aircraft shown in Figure 9.5, 200kt/250kt/M0.6: aircraft climbs from sea-level at 200 knot (IAS) to a chosen altitude, for example, 15,000 ft, then accelerates to 250 kt, and then climbs at 250 kt (IAS) to 23,500 ft, at which IAS = 250 kt and M of 0.6 represents the same TAS = 360 kt. This altitude of 23,500 ft is the cross-over altitude for the aircraft following the climb profile.

Increase M_{crit}

When an aircraft flies fast enough, and its free-stream Mach number reaches M_{crit}, a shockwave is able to form locally on aerofoil. The formation of a shockwave will cause drag increase and lift decrease, or even stall, or stability and control difficulties for transonic flights. Significant efforts have been made to increase M_{crit} of transonic flights, so an airplane can travel faster economically. Several design features will be discussed here to demonstrate ways to increase M_{crit} by following the aerodynamic basics.

Slimness

An aircraft with a slim wing will have a relatively higher M_{crit}. Figure 9.6 shows a thick aerofoil (a) and a thin aerofoil (b) in airflow. Examine the airflow on top of the aerofoil: if we treat the airflow as it passing through a half of convergent-divergent nozzle over the upper surface of the aerofoil, the airflow would increase its speed by following the principles of compressible airflow. The principles of 1-D compressible airflow, discussed in Chapter 7, tell us: $\frac{dA}{A} = (M^2 - 1)\frac{dv}{v}$, and $\frac{dA}{A} = \left(\frac{1 - M^2}{\gamma M^2}\right)\frac{dp}{p}$: the airspeed inside a convergent-divergent nozzle increases proportionally with the change of the area of flow path. For both thick and thin aerofoil, the airflow speed increases with the decrease of the area of the flow path before the most cambered/highest point (the "throat" – T in Figure 9.6 (a) and (b)) on the upper surface of the aerofoil. The airspeed increases from v_1, the free-stream airspeed to speed of sound, a_c, at the "throat". So the $M_{\text{crit}} = \frac{v_1}{a_1}$.

For a thick aerofoil, the change of the flow path "area", $|dA_{thick}|$, is greater than $|dA_{thin}|$, and at the "converging" part (before the "throat"), $dv_{thick} > dv_{thin}$. It means that the airspeed increases slower over a slim aerofoil than it does over a thick aerofoil. Since $v = a_c$ at the "throat" when $M_{fs} = M_{crit}$, v_{1thick}, the free-stream airspeed for the thick aerofoil will be less than v_{1thin}, the free-stream airspeed of the thin aerofoil, so the M_{crit} of the aerofoil with the slim design will be greater than M_{crit} of the thicker aerofoil. However, a thin aerofoil has a low lift coefficient, C_L, usually, as discussed in Chapter 4.

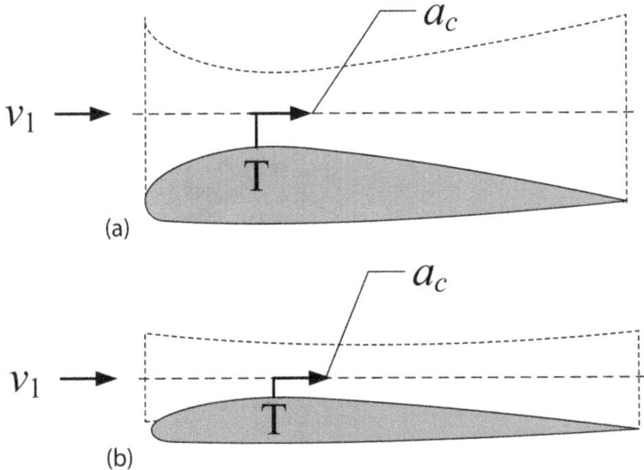

FIGURE 9.6
(a) Airflow over a transonic aerofoil; (b) Airflow over a thin transonic aerofoil.

Flat Leading Edge

The design of a flat leading edge can increase M_{crit} of an aerofoil for the same reason as that for the slim aerofoil, explained earlier. The outline shape of the upper surface of an aerofoil with the flat leading-edge design is similar to that of a slim aerofoil. Laminar flow aerofoil with a slim leading edge, and the aerofoil with supercritical section can have a higher M_{crit}. A supercritical aerofoil has a flattened top upper surface from the leading edge.

The thickness, t/c ratio, of the aerofoil with flat leading-edge design does not need to be as small as that of a slim aerofoil.

Sweepback

When air flows over an aerofoil along its chord increases to become sonic, the aircraft's Mach number reaches its critical Mach number M_{crit}. The aircraft with sweep back wings can have relatively higher M_{crit}.

M_{crit} is defined by free-stream airspeed and the free-stream speed of sound. For the same free-stream airspeed, the chordwise speed for a straight wing is the same as the free-stream airspeed shown in Figure 9.7 (a), while the chordwise speed is smaller than the free-stream speed for a sweep back wing as shown in Figure 9.7 (b).

However, it is the chordwise speed that determines whether/when a normal shockwave would occur over the aerofoil. Therefore, the free-stream airspeed, $v_{fs\text{-}straight}$, is less than the free-stream airspeed, $v_{fs\text{-}sweepback}$, when the chordwise speed reaches sonic. Λ is the sweep angle in Figure 9.7 (b), then the

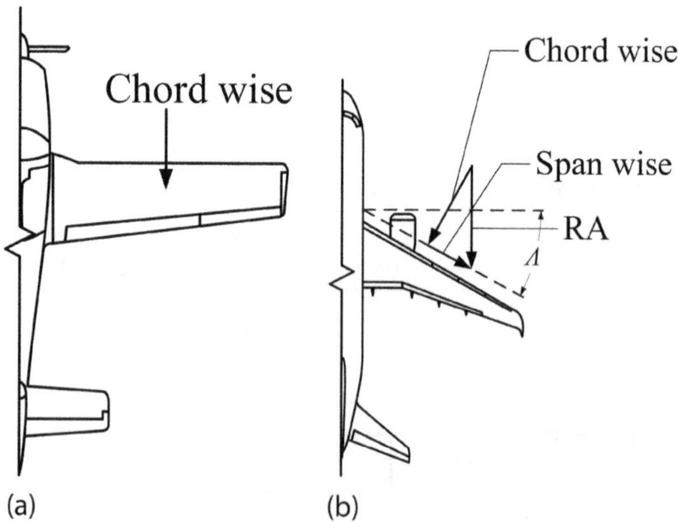

FIGURE 9.7
(a) Airspeed over a straight wing; (b) Airspeed over a sweep back wing.

sweep back wing's critical Mach number, $M_{crit\text{-sweepback}} \cdot \cos\Lambda = M_{crit\text{-straight}}$, the straight wing's critical Mach number. So the sweep back wing can increases its critical Mach number, M_{crit}. The greater the sweep angle, Λ is, the greater the critical Mach number, M_{crit}, will be.

Vortex Generators

Vortex generators are the small uneven devices on an aerofoil (similar to the ones shown in Figure 4.26). When airflow passes the vortex generators, microscopic energetic vortices are produced over the surface of the aerofoil. The vortices can delay the formation of shockwave, and delay the separation of the turbulent wake behind the normal shockwave if a normal shockwave has formed. So the aerodynamic difficulties caused by the normal shockwave on the aerofoil would be weakened and delayed. As the result of having vortex generators on top of aerofoil, the airplane can travel faster, which is equivalent to having a greater M_{crit}.

Transonic Aerofoils

Transonic airplanes experience the effects of shockwaves formed around their aerofoils. The main difficulties caused by shockwaves are the significant increase in drag, and decreases in lift, and shock stall. The transonic aerofoils are designed to overcome those difficulties, or to reduce the effects of shockwaves, and to delay those effects in their level cruises, before bow shockwaves are formed at their leading edges.

Low Thickness to Chord Ratio *t/c*

A thin design of wings is used for high-speed flights. The cross-section area of airflow path over an aerofoil increases gently if its *t/c* ratio is low. So a thin aerofoil will have higher M_{crit}, as discussed previously. For the same design, the cross-section area of airflow path over an aerofoil decreases gently as well after it passes the most cambered location. Therefore, the airflow speed over a thin aerofoil increases gently and smoothly, while the pressure over the aerofoil will decrease slowly over the whole aerofoil. This feature will delay the formation of shockwaves, and the shockwave would be relatively weak if it should be formed.

The intensity of a shockwave depends on the Mach number before the shockwave. The higher the Mach number before the shockwave is, the stronger the shockwave will be. A strong shockwave produces a high shock drag; the turbulent wake behind the shockwave is more unstable; and the Mach buffet and shock stall can occur easily. Therefore, a wing with a low *t/c* ratio

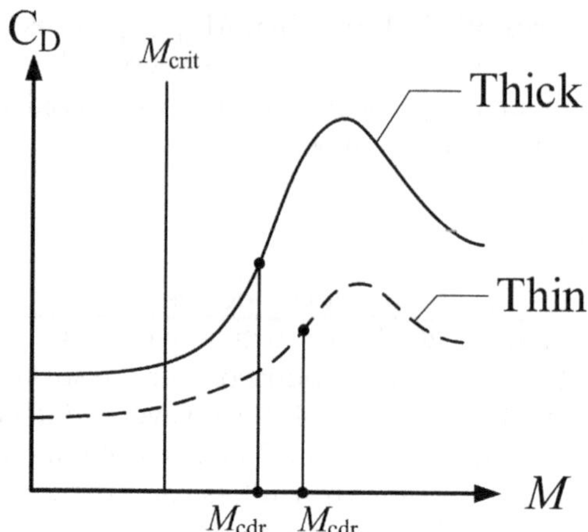

FIGURE 9.8
Comparison of C_D between a thin and a thicker aerofoil.

will experience a low shock drag, and produce a higher M_{cdr}, the critical drag rise Mach number (or called M_{dd}: drag-divergence Mach number), as shown in Figure 9.8. Because the Mach number over the thin wing will not increase quickly, the turbulent wake behind the shockwaves will be less violent, so the aircraft with thin aerofoil travels at high speed with a low shock drag and better stability. It is possible that the buffet margin of an aircraft with thin wings is greater than that of the aircraft with thick wings.

However, because the pressure decreases slowly over a thin wing, it will produce less lift than the aerofoil with a greater thickness to chord ratio, *t/c*, especially at lower air speeds. So the lift coefficient of a thin aerofoil is relatively low.

Figure 9.9 shows the comparison of lift coefficient C_L between a thin aerofoil and relatively thick aerofoil; the lift coefficient of the thinner one is much

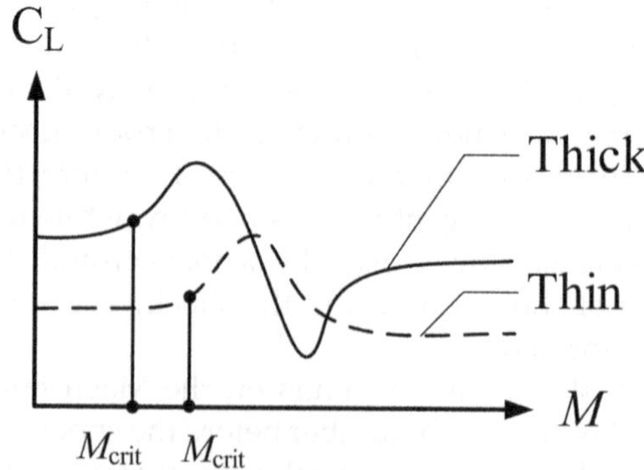

FIGURE 9.9
Comparison of C_L between a thin and a thicker aerofoil.

lower, in particular, at low speeds, but the critical Mach number M_{crit} of the thin one is higher (discussed previously in this chapter).

Supercritical Aerofoil

A supercritical aerofoil has a specially designed cross-section for transonic airplanes. This supercritical section is designed to reduce shock drag, and to delay shock stall.

Figure 9.10 shows the supercritical section. In comparison with an ordinary aerofoil cross-section, the features of the supercritical section are:

- Flattened-up surface at leading edge with a modest *t/c*, thickness to chord ratio. This feature ensures that the increase of airflow speed over the aerofoil is gentle, and smooth like that over a thin aerofoil, so its critical Mach number, M_{crit}, is higher than that of a conventionally shaped aerofoil. Because its thickness is not low, it does not have the disadvantage of low lift coefficient, as a thin aerofoil does.
- This feature increases M_{cdr} as well, since the airflow speed will not increase significantly, the formation of the shockwave on top of the aerofoil will be delayed, and the intensity of the normal shockwave will be reduced when it forms. So the significant increase in shock drag will occur when M_{fs} is relatively high and the aircraft can cruise at higher subsonic speeds.
- Another special feature of the supercritical aerofoil is the *reflex camber*, as shown in Figure 9.10. The camber line is reflected upward at the rare part of the aerofoil. It improves the lift production at the rear part of the aerofoil. When a normal shockwave just forms at lower surface, and the airflow behind the shockwave becomes subsonic, the reflex camber could cause the area of the flow path at the rear part of the lower surface increase to form a relatively high air pressure region to maintain lift production.
- Because the supercritical aerofoil can reduce shock drag and improve the lift coefficient, it can make the aircraft operate more efficiently. The sweep angle and the wing span can be reduced if the sweepback wing uses the supercritical section design.

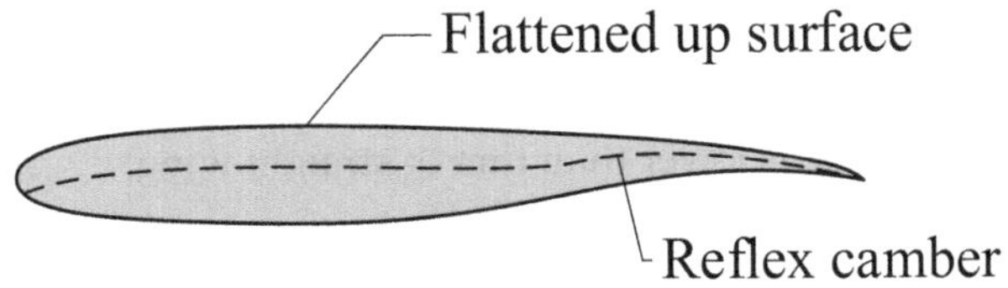

FIGURE 9.10
Supercritical aerofoil.

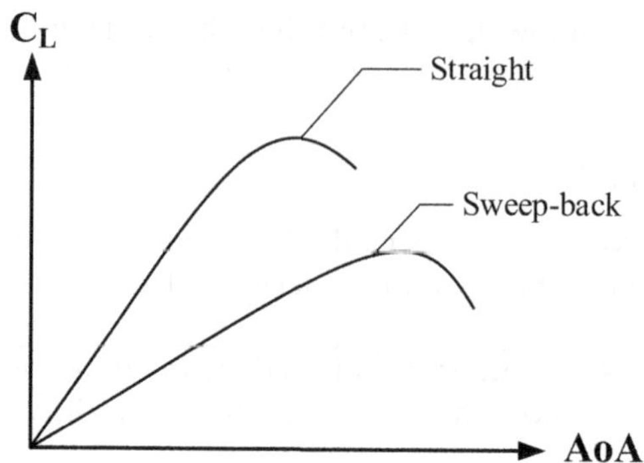

FIGURE 9.11
Comparison of C_L between a straight wing and a sweep back wing.

Sweepback Wings

Sweepback wings are popular designs for transonic airplanes. As we discussed in Chapter 5, sweepback wings provide positive longitudinal, lateral and direction stability in flight, and can increase the M_{crit} of transonic aircraft as mentioned above.

However, there are some issues related to actual performance of sweepback wing and the implementing its ideal principle of the design. It is important to understand those issues and learn the methods to overcome the disadvantages caused by those issues in order to know the aircraft you will fly better, and to control it in flight wisely.

The characteristics of sweepback wings:

- Sweepback wings have a positive stability in any direction. The recovering pitch, roll, and yaw moment can be produced by a sweepback wing if the aircraft is in a pitch, roll, and yaw disturbance, respectively.
- Sweepback wings produce higher M_{crit} as shown earlier, and the higher sweep angle is the higher M_{crit} will be.
- For the same reason, M_{cdr} is higher for an aircraft with sweepback wings. This means that the sweepback wing design can delay the shockwave formation and the turbulent wake separation behind the shockwave. So sweepback wings produce relatively low shock drag.
- The stall angle of attack for a sweepback wing is higher than that of straight wing, as shown in Figure 9.11. But the lift coefficient of sweep back wing is lower than that of straight wing.
- The transonic aircraft with sweepback wings can operate at relatively high economical cruising speeds, as its drag coefficient C_D is relatively low.

- Airflow around sweepback wing tip departs from the free-stream direction. Because air is a viscous fluid, the local airflow at the wingtip changes its direction when it flows around the sweep wingtip, as shown in Figure 9.12. The effective sweep angle at the wingtip is smaller than the sweep angle of the wing, so the effective M_{crit} is reduced from the theoretic M_{crit}.
- The thickness of sweep wing is not uniform along its wing span. The thickness reduces gradually from wing root to wingtip. This feature makes the airflow over the wing moving toward to the wing root. As the result, a series of compression waves are formed close to the wing root (the dashed lines in Figure 9.12), even a shockwave can be built from the compression waves. Therefore, drag will be increased due to those compression waves.
- There is a spanwise component of airflow over a sweep wing, as shown in Figure 9.7 (b). As the result of the spanwise flow, the boundary layer over the sweepback wing is getting thicker toward to the wingtip. The thickened wingtip boundary layer will encourage the formation of a large wingtip vortex. The wingtip vortex is so large, when the wing's sweep back angle increases, that it can start from the leading edge of the wing, which is called a *"ram horn"* vortex, as shown in Figure 9.13 (a). The large vortex induces more downwash, which, in turn, leads to higher induced drag.

  The thickened wingtip boundary layer can cause the boundary layer separation at the wing tip easily, which leads to *tip stall.* The center of pressure, CP, will move forward when tip stall takes place, shown in Figure 9.13 (b), and it will decrease the restoring pitching moment if the aircraft is in a pitching disturbance. The "wash-out" design of a sweep back wing will reduce/prevent tip stall.
- When shockwave forms over a transonic sweepback wing, the turbulent wake behind the shockwave is situated further back close to its tail, and it can shadow T-tail (if a plane has configured with a T-tail). The separation of the turbulent wake causes the loss of lift, and can introduce instability in pitching. The turbulent wake can shadow the tail plane when it becomes violently separating from the aerofoil surface. The tail plane cannot work effectively in the shadow – the restoring function of the tail plane is compromised, in particular, when the aircraft in the disturbance caused by the separation. This situation that the aircraft is going to stall but the tail plane cannot help is called *deep stall.* A high T-tail design can be used to set the tail plane away from the turbulent wake and to avoid deep stall.

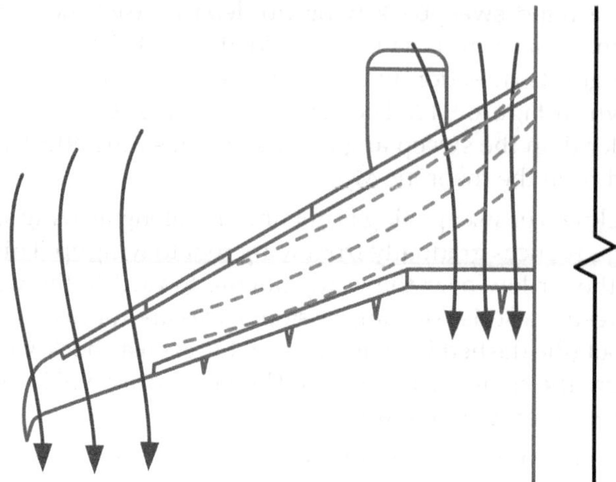

FIGURE 9.12
Direction change of airflow over a sweep back wing.

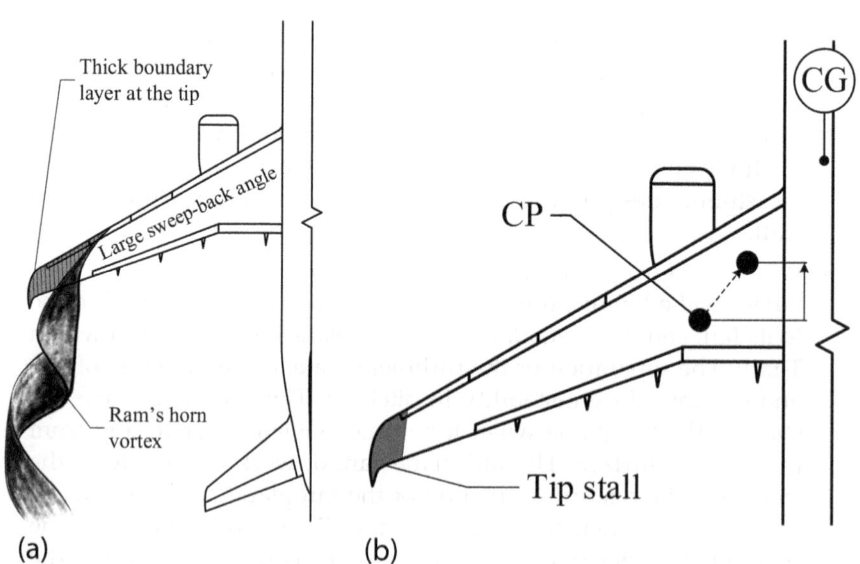

FIGURE 9.13
(a) Thickened boundary layer at the tip of a sweep back wing causing "ram horn" vortex; (b) Change of CP caused by sweep back wing tip stall.

Devices to Delay Shock Stall

It is importance for transonic aerofoil to prevent shock stall. Figure 9.14 and Figure 9.15 show several devices used on transonic wings to delay or prevent shock stall.

Wing Fences

Wing fences, as shown in Figure 9.14, can be featured on a wing close to the wingtip area to prevent a large wingtip vortex to develop into a "ram horn" vortex, which causes high induced drag and wingtip stall. Wing fences can also interrupt the rear shockwave over a sweepback wing developing widely, and then reduce the turbulent wake separation behind the shockwave.

FIGURE 9.14
Wing fence and vortex generators.

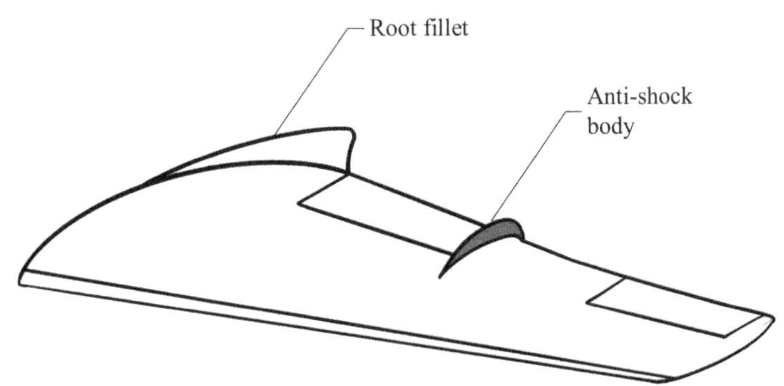

FIGURE 9.15
Some anti-shock bodies.

Vortex Generators

Vortex generators are the common feature on a transonic wing, shown in Figure 4.26, and Figure 9.14. They normally are installed near the leading edge of a wing. Vortex generators produce streets of microscopic vertices over the wing within the boundary layer, to transfer more kinetic energy into the boundary layer. With a higher level of kinetic energy, boundary layer separation will be delayed, so the stall caused by the boundary layer separation will be delayed or prevented. When the free-stream Mach number excesses its M_{crit}, the vortex generators could destabilize the formation of the incident normal shockwave over the wing. It delays the formation of shockwave, delays the separation of turbulent wake behind the shockwave, and then delays the shock stall.

Anti-Shock Body

A design of anti-shock body is called *Küchemann carrots,* which are streamlined pod shape bodies, as shown in Figure 9.15. They are added to the leading or trailing edge of an aerodynamic surface, starting from near the point of maximum thickness, or most camber and extending beyond the trailing edge of a wing. They may prevent the local airspeed increases much further, so the local Mach number would not be very high, and the intensity shockwave on the aerofoil would not be high. So Mach buffet will be improved.

- The anti-shock bodies can reduce interference of flow streams from different parts of the aircraft, so it will reduce the buffeting of the wing in the transonic range and reduce interference drag.
- The local airflow caused by those anti-shock bodies interrupts the shockwave when it moves toward the trailing edge when free-stream Mach number increases. Therefore, Mach buffet will be reduced, and the shock stall will be delayed.
- Because those anti-shock bodies produce airflow to delay the shockwave, and the delayed shock stall, they can reduce transonic wave drag.

Area Rule

The *area rule* of design of transonic aircraft is that to achieve the minimum transonic drag rise, the cross-sectional area of the whole aircraft should increase and decrease smoothly from nose to tail. Figure 9.16 (a) and (b) shows that the difference of the designs with and without area rule, and Figure 9.17 shows the difference of the transonic drag coefficient the area rule can make.

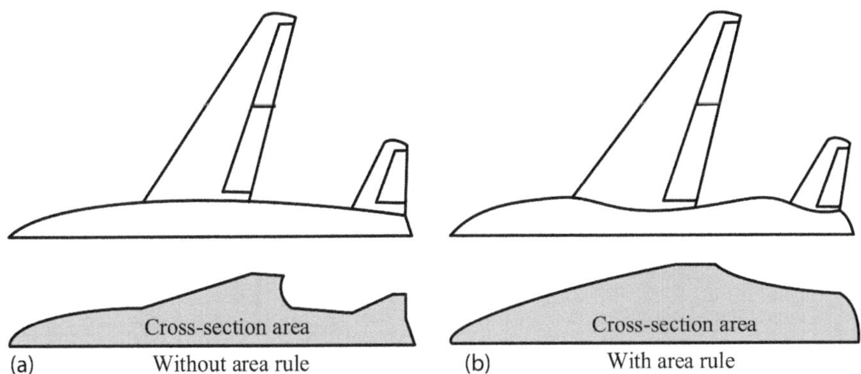

FIGURE 9.16
(a) Cross-section profiles with and without area rule; (b) Comparison of C_D with and without area rule.

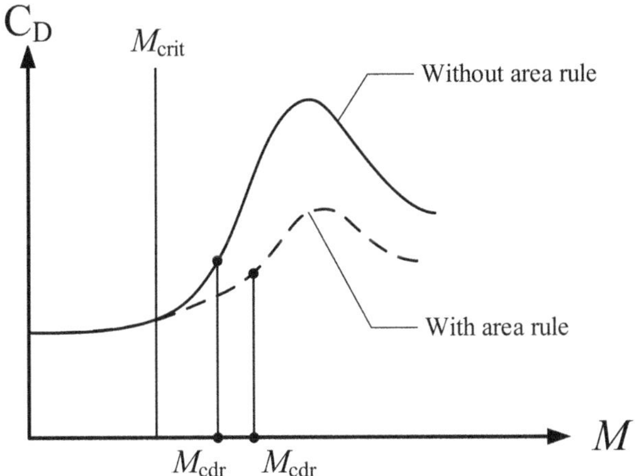

FIGURE 9.17
Comparison of drag coefficient with and without area rule.

Exercises

1. Two climb profiles 225 kt/280 kt/M0.75 and 250 kt/300 kt/M0.75: to which profile is the cross-over altitude higher?
2. What is the difference between coffin corner and cross-over altitude?
3. Why does supercritical aerofoil produce less transonic drag?

10

Supersonic Waves

The change or the disturbance in air property always trails behind the moving air particles when airflow is supersonic. As a result, the characteristics of aerodynamics of the airflow change so much that they become fundamentally different from those of air in subsonic flow. The changes of air property in supersonic flow form special waves that are very important in the design of a supersonic airplane. In this chapter, the two forms of supersonic waves, the oblique shock wave and the expansion wave, will be analyzed from the formation of those waves to the effects of the waves and their aerodynamic features.

The supersonic waves are formed within a supersonic flow in isolation, i.e., there is no heat exchange to or from outside the airflow system, in an adiabatic process. However, those two types of supersonic waves are formed in different adiabatic processes: one is reversible and the other is irreversible.

Two Examples of Reversible and Irreversible Adiabatic Processes (Supersonic)

Reversible

The airflow in a convergent-divergent nozzle, shown in Figure 7.5 and discussed in Chapter 7, is in a reversible adiabatic process, which means that the property of the airflow changes in the nozzle, and the property of the airflow can be reversed back to the original state completely. In this process, the pressure at the inlet p_1 is much higher than the pressure at the outlet p_2, when a compressible subsonic airflow passes through this convergent-divergent nozzle from left to right, and becomes supersonic at the exit. The airspeed and air property change gradually throughout the nozzle. If the process described here is assumed frictionless, it is a reversible process. There is no energy loss and no heat transfer, and then, the process of the airflow through this convergent-divergent nozzle is known as the reversible adiabatic process, also called *isentropic process.*

Therefore, the stagnation temperature T_o of the airflow described above is the same from the left end to the right end of the channel determined by the Energy equation (2.37), and the stagnation pressure p_o is the same from the left end to the right end of the channel, as well. The p_o can be calculated by formulae (2.29).

Irreversible

The airflow passes through a normal shockwave, as shown in Figure 7.6 and discussed in Chapter 7, is an irreversible adiabatic process. In this process, the compressible supersonic flow passes through a normal shockwave, and it becomes subsonic. There is no heat transfer during this process, but the normal shockwave consumes the kinetic energy of the airflow. After the airflow passes the shockwave, the air pressure, density and temperature increase suddenly. The supersonic airflow suddenly becomes subsonic after the shockwave. This sudden change is not reversible because a subsonic airflow could not pass through a normal shockwave, and suddenly became supersonic. Supersonic airflow in a free-stream travel to the front of a normal shockwave and the subsonic airflow travels away after the shockwave can both be reversible, but the process of airflow passing through the shockwave is an *irreversible* adiabatic process.

The stagnation temperature T_{o1} before the shockwave and T_{o2} after the shockwave are the same, $T_{o1} = T_{o2}$, since there is no heat transfer. The stagnation temperature before (subscript "1") and after (subscript "2") the shockwave can be expressed by the Energy equation (7.7), respectively:

$$\frac{\gamma R_M}{\gamma - 1} T_{o1} = \frac{\gamma R_M}{\gamma - 1} T_1 + \frac{v_1^2}{2}, \quad \text{then} \quad \frac{T_{o1}}{T_1} = 1 + \frac{(\gamma - 1)}{2} M_1^2; \tag{10.1}$$

$$\frac{\gamma R_M}{\gamma - 1} T_{o2} = \frac{\gamma R_M}{\gamma - 1} T_2 + \frac{v_2^2}{2}, \quad \text{then} \quad \frac{T_{o2}}{T_2} = 1 + \frac{(\gamma - 1)}{2} M_2^2; \tag{10.2}$$

The stagnation pressure before the shockwave according to formulae (2.29):

$$p_{o1} = p_1 \left(\frac{T_{o1}}{T_1} \right)^{\frac{\gamma}{\gamma - 1}} \tag{10.3}$$

The stagnation pressure after the shockwave according to formulae (2.29):

$$p_{o2} = p_2 \left(\frac{T_{o2}}{T_2} \right)^{\frac{\gamma}{\gamma - 1}} \tag{10.4}$$

Take a ratio of Equation (10.4) to Equation (10.3), and $T_{o1} = T_{o2}$:

$$\frac{p_{o2}}{p_{o1}} = \frac{p_2}{p_1} \left(\frac{T_1}{T_2} \right)^{\frac{\gamma}{\gamma - 1}} = \frac{p_2}{p_1} \left[\frac{2 + (\gamma - 1) M_2^2}{2 + (\gamma - 1) M_1^2} \right]^{\frac{\gamma}{\gamma - 1}} \tag{10.5}$$

It shows that $p_{o2} \neq p_{o1}$. It is also can be found that $p_{o2} < p_{o1}$. The ratio of $\frac{p_2}{p_1}$ in Equation (10.5) increases, when M_1 increases, as discussed in Chapter 7, and the ratio of $\left(\frac{T_1}{T_2}\right)^{\frac{\gamma}{\gamma-1}}$ decreases much faster with the increase of M_1. As the result, the ratio of $\frac{p_{o2}}{p_{o1}}$ is always less than "1", i.e. the stagnation pressure (also known as "total pressure") of the airflow will be decreased after it passes a normal shockwave.

If we divide both sides of the Equation (10.4) by p_1, the static pressure before the shockwave:

$$\frac{p_{o2}}{p_1} = \frac{p_2}{p_1}\left(\frac{T_{o2}}{T_2}\right)^{\frac{\gamma}{\gamma-1}} \tag{10.6}$$

Substitute Equation (7.37) for $\frac{p_2}{p_1}$, Equation (10.2) for $\frac{T_{o2}}{T_2}$, and Equation (7.33) for M_2^2 into (10.6). Rearranging the equation, and then the Rayleigh supersonic Pitot equation (6.21) will be obtained:

$$\frac{p_{o2}}{p_1} = \frac{\left(\frac{\gamma+1}{2}M^2\right)^{\frac{\gamma}{\gamma-1}}}{\left[\frac{2\gamma M^2-(\gamma-1)}{\gamma+1}\right]^{\frac{1}{\gamma-1}}}$$

Oblique Shockwaves

When a supersonic airflow encounters compression pressure disturbance, a shockwave will be formed. For example, a supersonic airflow passes over a concave corner, as shown in Figure 10.1. The airflow at the corner is compressed, and a series of consecutive pressure waves (Mach waves in supersonic flow) are generated and accumulated at the corner, then, a shockwave is formed at the concave corner with an angle to the incoming airflow. This angle between the shockwave and the direction of the incoming airflow is less than 90°. This shockwave is called *oblique shockwave*. A normal shockwave can be seen as a special form of an oblique shockwave when the angle is 90°.

Oblique shockwaves can be observed at the leading edge of a supersonic aircraft, as shown in Figure 10.2 (a), and inside a bent duct/pipe with supersonic airflow as shown in Figure 10.2 (b).

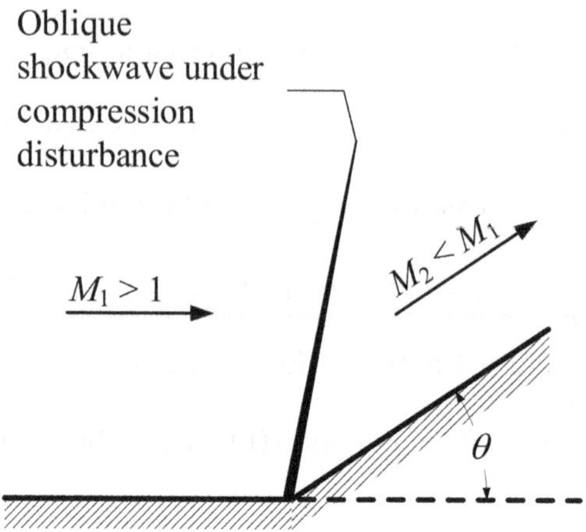

FIGURE 10.1
Oblique shockwave on a deflected surface.

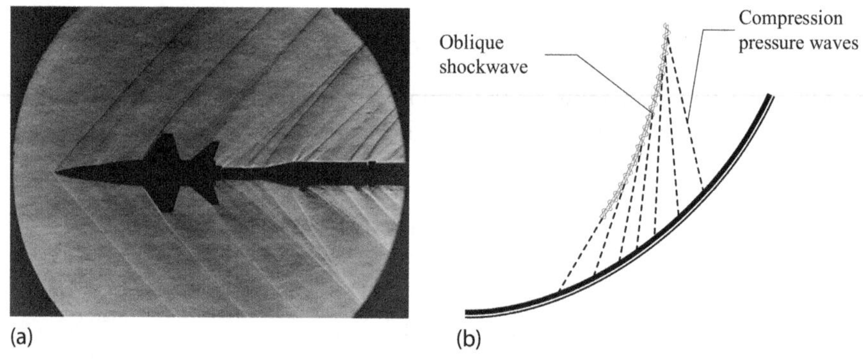

FIGURE 10.2
(a) Oblique shockwaves on a supersonic aircraft; (b) Oblique shockwave with supersonic gas flow inside a bent duct.

Air Properties Before and After an Oblique Shockwave

The air properties after an oblique shockwave are very important to lift, drag, and the stability of flight control of a supersonic aircraft, and the design of effective jet engines, rockets, and other airspace objects at supersonic speeds.

An oblique shockwave is a compression supersonic wave, so the air pressure increases when it passes through an oblique shockwave. A normal shockwave is a special case of oblique shockwave. The relationship of air properties before and after a normal shockwave will be used in deriving the relationships of air properties before and after an oblique shockwave. The air movement across an oblique shockwave can be analyzed in two orthogonal directions: parallel to the shockwave and perpendicular to the shockwave, as shown in Figure 10.3.

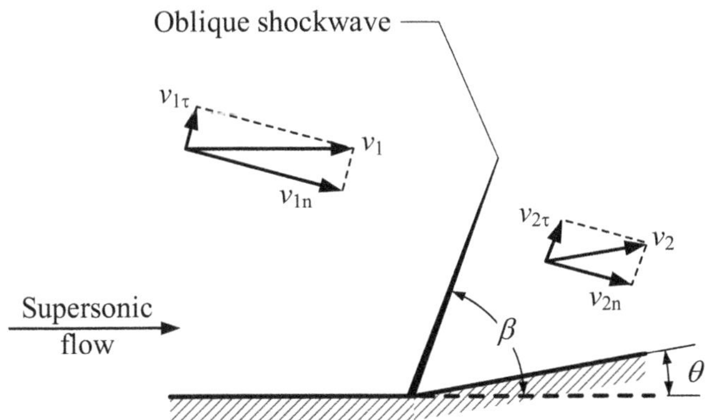

FIGURE 10.3
Components of air velocity before and after an oblique shockwave.

A supersonic airflow travels over a deflected surface (a concave corner) and an oblique shockwave is formed at the concave corner in Figure 10.3. The surface deflected upward by an angle of θ. θ is called *deflection angle*. β, which marks the angle between the oblique shockwave and the direction of the airflow before the shockwave, is called the *shock angle*.

The supersonic airflow changes its direction after it passes through this oblique shockwave. It then flows parallel to the deflected surface. The air velocity before the shockwave is v_1, and the velocity after the shockwave is v_2. The subscript, τ, indicates the direction parallel to the shockwave, and the subscript, n, indicates the direction perpendicular to the shockwave. The air velocity can be divided into two components in those two directions: $v_{1\tau}$ and v_{1n} are the air velocity components parallel and perpendicular to the shockwave, respectively, before the shockwave; and $v_{2\tau}$ and v_{2n} are the air velocity components parallel and perpendicular to the shockwave after the shockwave as shown in Figure 10.3. Applying trigonometry rules, we obtain:

$$v_{1\tau} = v_1 \cos\beta \tag{10.7}$$

$$v_{1n} = v_1 \sin\beta \tag{10.8}$$

$$v_{2\tau} = v_2 \cos(\beta - \theta) \tag{10.9}$$

$$v_{2n} = v_2 \sin(\beta - \theta) \tag{10.10}$$

and the corresponding Mach numbers, since the air temperature/speed of sound is the same in both directions:

$$M_{1\tau} = M_1 \cos\beta \tag{10.11}$$

$$M_{1n} = M_1 \sin\beta \tag{10.12}$$

$$M_{2\tau} = M_2 \cos(\beta - \theta) \tag{10.13}$$

$$M_{2n} = M_2 \sin(\beta - \theta) \tag{10.14}$$

In τ direction: there is no slippage in airflow, which means that the velocity components in this direction at both sides of the shockwave are the same: $v_{1\tau} = v_{2\tau}$.

In n direction: the air passes the shockwave, and the velocity components are perpendicular to the shockwave, as if airflow passes through a normal shockwave, so according to the analysis in Equation (7.32) in Chapter 7, we can obtain: $v_{1n} \cdot v_{2n} = a_c^2$, or $M_{1n}^* M_{1n}^* = 1$.

The Mach number will decrease to <1 (subsonic), after a normal shockwave. The M_{2n} should become less than 1, and can be calculated by Equation (7.33):

$$M_{2n}^2 = \frac{2 + (\gamma - 1) M_{1n}^2}{2\gamma M_{1n}^2 - (\gamma - 1)} \tag{10.15}$$

Substitute Equations (10.12) and (10.14) into Equation (10.15) to obtain the Mach number after an oblique shockwave, M_2:

$$M_2^2 = \frac{2 + (\gamma - 1) M_1^2 \sin^2\beta}{2\gamma M_1^2 \sin^2\beta - (\gamma - 1)} \cdot \frac{1}{\sin^2(\beta - \theta)} \tag{10.16}$$

Please note that although the oblique shockwave, M_{2n} is smaller than "1", M_2 can be still greater than "1" depending on the shock angle, β.

Air properties change instantly when it flows across an oblique shockwave. v_τ does not contribute to this change; only v_n does. So, the change of air property is determined by the change of velocity component, or Mach number in n direction, as if the airflow passed through a normal shockwave. Equations (7.35), (7.36), and (7.37) can be used to determine the change of air density, pressure, and temperature after this oblique shockwave, replacing M_1 with M_{1n}:

$$\frac{\rho_2}{\rho_1} = \frac{(\gamma + 1) M_{1n}^2}{2 + (\gamma - 1) M_{1n}^2} \tag{10.17}$$

$$\frac{p_2}{p_1} = \frac{2\gamma}{(\gamma + 1)} M_{1n}^2 - \frac{\gamma - 1}{\gamma + 1} \tag{10.18}$$

$$\frac{T_2}{T_1} = 1 + \frac{2(\gamma - 1)}{(\gamma + 1)^2} \frac{(\gamma M_{1n}^2 + 1)}{M_{1n}^2} (M_{1n}^2 - 1) \tag{10.19}$$

Substitute Equation (10.12) into Equations (10.17), (10.18), and (10.19) to obtain the relationships of air properties before and after an oblique shockwave:

$$\frac{\rho_2}{\rho_1} = \frac{(\gamma + 1) M_1^2 \sin^2 \beta}{2 + (\gamma - 1) M_1^2 \sin^2 \beta} \tag{10.20}$$

$$\frac{p_2}{p_1} = \frac{2\gamma}{(\gamma + 1)} M_1^2 \sin^2 \beta - \frac{\gamma - 1}{\gamma + 1} \tag{10.21}$$

$$\frac{T_2}{T_1} = 1 + \frac{2(\gamma - 1)}{(\gamma + 1)^2} \frac{(\gamma M_1^2 \sin^2 \beta + 1)}{M_1^2 \sin^2 \beta} (M_1^2 \sin^2 \beta - 1) \tag{10.22}$$

Equations (10.16), (10.20), (10.21), and (10.22) describe the speed change and the air property difference when a supersonic airflow passes through an oblique shockwave.

Example 10.1

The Mach number M_1 of an incoming airflow is 1.5 under sea level conditions. It flows over a concave corner as shown in Figure 10.3, and an oblique is formed at the corner. The deflection angle θ of the surface is 10°, and the shock angle of the oblique shockwave is $\beta = 56°$.

1. Find out the Mach number M_2, air pressure p_2, density ρ_2, and temperature T_2 behind the shockwave.
2. What is M_{2n}?
3. Is the airflow after this shockwave subsonic?
4. What is the air speed after the shockwave?

SOLUTION

Air at sea level conditions: $p_1 = 1.013 \times 10^5$ Pa, $\rho_1 = 1.225$ kg/m³, and $T_1 = 288$ K.

1. Use Equation (10.16) to calculate M_2 ($\gamma = 1.4$ for air):

$$M_2^2 = \frac{2 + (\gamma - 1) M_1^2 \sin^2 \beta}{2\gamma M_1^2 \sin^2 \beta - (\gamma - 1)} \cdot \frac{1}{\sin^2(\beta - \theta)} = \frac{2 + (1.4 - 1) \times 1.5^2 \times (\sin 56°)^2}{2 \times 1.4 \times 1.5^2 \times (\sin 56°)^2 - 0.4}$$

$$\times \frac{1}{(\sin(56° - 10))^2}$$

$$M_2 = \sqrt{\frac{2.62}{3.93} \times \frac{1}{0.52}} = 1.13.$$

Use Equations (10.20), (10.21), and (10.22) to calculate air density ρ_2, pressure p_2, and temperature T_2:

$$\frac{\rho_2}{\rho_1} = \frac{(1.4+1)\times 1.5^2 \times \sin^2 56°}{2+(1.4-1)\times 1.5^2 \times \sin^2 56°} = 1.42;\ \rho_2 = 1.42\times 1.225 = 1.736\ \text{kgm}^{-3}.$$

$$\frac{p_2}{p_1} = \frac{2\times 1.4}{(1.4+1)}\times 1.5^2 \sin^2 56° - \frac{1.4-1}{1.4+1} = 1.64,\ \text{and}\ p_2 = 1.64\times 1.013\times 10^5$$

$$= 1.659\times 10^5\ \text{pa};$$

$$\frac{T_2}{T_1} = 1+\frac{2\times 0.4}{2.4^2}\times\frac{(1.4\times 1.5^2\times \sin^2 56°+1)}{1.5^2 \sin^2 56^2}(1.5^2 \sin^2 56° - 1) = 1.16$$

and $T_2 = 1.16\times 288 = 332$ K.

2. Use Equation (10.14) to calculate M_{2n}, and $M_2 = 1.13$:

$$M_{2n} = M_2 \sin(\beta-\theta) = 1.13\times \sin(56°-10°) = 0.81.$$

3. No, the airflow after the oblique shockwave is supersonic, because $M_2 = 1.13 > 1$.
4. The Mach number $M_2 = v_2/a_2$, so

$$v_2 = M_2 a_2 = 1.13\times\sqrt{\gamma R_m T_2} = 1.13\times\sqrt{1.4\times 287\times 332} = 412.72\ \text{ms}^{-1}.$$

The airspeed after the shockwave is 412.72 ms^{-1}.

θ–β–M Equation

Airflow passes through an oblique shockwave, and there is no velocity difference in τ direction. Applying the Continuity equation in n direction, the area, at which the airflow passes through the oblique shockwave, is the same to the airflow before and after the shockwave, so the Continuity equation can be written as:

$$\rho_1 v_{1n} = \rho_2 v_{2n} \tag{10.23}$$

In Figure 10.3, it is shown that:

$$v_{1n} = v_{1\tau} \tan\beta \tag{10.24}$$

$$v_{2n} = v_{2\tau} \tan(\beta-\theta) \tag{10.25}$$

From Equation (10.23), the following can be obtained using Equations (10.24) and (10.25):

$$\frac{v_{2n}}{v_{1n}} = \frac{\tan(\beta-\theta)}{\tan\beta} = \frac{\rho_1}{\rho_2} \tag{10.26}$$

Substitute Equation (10.20) into Equation (10.26):

$$\tan(\beta - \theta) = \frac{2 + (\gamma - 1)M_1^2 \sin^2 \beta}{(\gamma + 1)M_1^2 \sin \beta \cos \beta} \tag{10.27}$$

Apply the rules in trigonometry to expand $\tan(\beta - \theta)$, and it can be found that:

$$\tan \theta = 2 \cot \beta \frac{M_1^2 \sin^2 \beta - 1}{M_1^2(\gamma + \cos 2\beta) + 2} \tag{10.28}$$

Equation (10.28) expresses the relationship between incoming Mach number, M, deflection angle θ and the shock angle β, so Equation (10.28) is called the *θ–β–M Equation*.

This equation tells us that for each given Mach number of incoming air flow, M_1, there is a relationship between the deflection angle θ, and shock angle β: when a flow passes a deflected surface, the shock angle, β, of the oblique shockwave depends on the deflection angle, θ. Figure 10.4 is the *θ–β–M* diagram made from *θ–β–M* Equation (10.28), when $M_1 = 1$, 1.1,

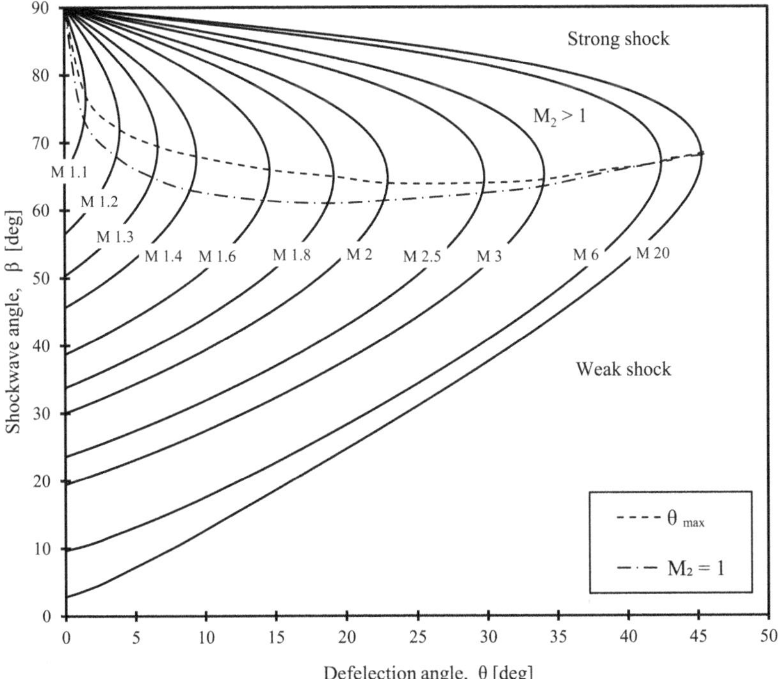

FIGURE 10.4
θ–β–M diagram.

1.2, 1.3, … This means that Mach number is constant on each solid line. The horizontal axis is deflection angle θ and the vertical axis is shock angle β in the θ–β–M diagram.

Each solid line in Figure 10.4 concaves toward to the β axis. At the turning point, the "nose" of each solid line, there is the corresponding deflection angle θ, and this θ is the maximum deflection angle θ_{max} for its M_1. Joining all the points at the "nose" forms a *maximum deflection angle line,* which is the upper dash-line in Figure 10.4. We can find what the maximum deflection angle of each Mach number is along this line. For examples, if $M_1 = 1.6$, $\theta_{max} = 14.5°$; and if $M_1 = 2.5$, $\theta_{max} \approx 30°$.

If the deflection angle, θ, of a surface is smaller than the θ_{max} at one M_1, there are two possible value of β, which means that there are two possible oblique shockwaves: the smaller one of those two β, indicates the oblique shockwave is a weak shock. The airflow after this oblique shockwave is supersonic, $M_2 > 1$. On the other hand, the greater one of those two β indicates the oblique shockwave formed is a strong shock, $M_2 < 1$, and the airflow after this oblique shockwave is subsonic. For example, if $M_1 = 1.6$, $\theta = 10° < \theta_{max} = 14.5°$, it is can be found in Figure 10.4 that $\beta_{weak} = 52°$, and $\beta_{strong} = 79°$. Therefore, the lower part of each solid line represents weak shocks, while the upper part represents strong shocks. There is a special point between the lower and the upper parts of each solid line, which indicates the airflow is sonic after the oblique shockwave: $M_2 = 1$. The lower dash-line (the *critical Mach line*) in Figure 10.4 represents the oblique shockwaves, after which the Mach number of the airflow is, $M_2 = 1$. For example, if $M_1 = 1.6$, and $\theta = 14°$, the airflow after the oblique shockwave will be sonic: $M_2 = 1$, and the shock angle β is approximate 62°.

When incoming Mach number is large, for example, $M_1 > 2.5$, the two dash lines, maximum deflection angle and the critical Mach line, lie very close to each other. Therefore, it can be assumed that airflow is sonic $M_2 = 1$ behind the oblique shockwave when the deflection angle is equal to the maximum deflection angle if the Mach number before an oblique shockwave is relatively high.

If the deflection angle is "0", i.e. a flat surface, the strong shock angle will be $\beta_{strong} = 90°$, and it is a *normal shockwave,* no matter what is the incoming Mach number, M_1; but the weak shock angle depends on the value of the incoming Mach number M_1, $\beta_{weak} = \sin^{-1}\left(\dfrac{1}{M_1}\right)$, same as Mach angle.

In general, strong shocks occur in an internal supersonic flow, for example, inside a turbine or at the exit of a jet engine, and weak shocks occur in an external supersonic flow, for example, the airflow over aerofoil, or fuselage of aircraft, or a traveling bullet.

If the deflection angle is greater than the maximum deflection angle at a M_1, $\theta > \theta_{max}$, the oblique shockwave will be detached from the leading edge of the deflected corner, as shown in Figure 10.5. For example, if $M_1 = 1.6$, and $\theta = 20° > \theta_{max} = 14.5°$, the oblique shockwave will be detached from the leading edge. For example, a bow shockwave is formed and detached at the leading edge of a transonic aerofoil, when the free-stream Mach number of the

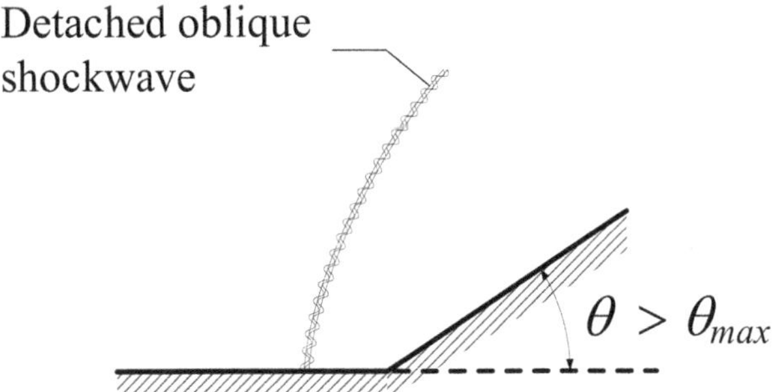

FIGURE 10.5
A detached oblique shockwave when $\theta > \theta_{max}$.

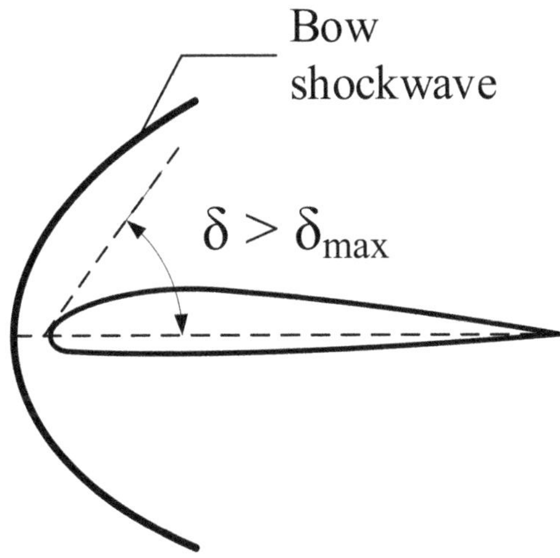

FIGURE 10.6
A detached bow shockwave at the leading edge of a transonic aerofoil.

aircraft reaches the detachment Mach number, i.e. $M_{fs} = M_{det}$. Generally, M_{det} is approximately, 1.2, and the corresponding θ_{max} is about 4°; the deflection angle at the leading edge of a transonic aerofoil is much greater than 4°, as shown in Figure 10.6.

Appendix II presents a large version of the θ–β–M diagram.

Example 10.2

Airflow is deflected by 10° along a surface. The free-stream Mach number is $M_1 = 1.6$ at sea level conditions. An oblique shockwave is formed.

a) What is the free-stream airspeed?
b) Is the oblique shockwave attached or detached at the leading edge of the deflected surface?

c) If the shockwave is a weak shock, find the Mach number, airspeed, and air properties after the oblique shockwave, i.e., M_2, v_2, p_2, ρ_2, and T_2.
d) What is the weak shock angle if M_1 is increased to 1.8?
e) What is β and θ, if $M_2=1$, when $M_1=1.8$?

SOLUTION

At the sea level conditions: $p_1=1.013\times10^5$ Pa, $\rho_1=1.225$ kg/m^3, and $T_1=288$ K.

a) The free-stream airspeed $v_1=M_1a_1$, where a_1 is the free-stream speed of sound.

$$v_1 = M_1\sqrt{\gamma R_m T_1} = 1.6\times\sqrt{1.4\times287\times288} = 544.28\ (\text{m/s}).$$

b) The maximum deflection angle when $M_1=1.6$ in Appendix I is 14.5°, and $\theta=10°<14.5°$. Therefore, the oblique shockwave is attached to the deflected surface.
c) The oblique shockwave is a weak shock, the shock angle is $\beta_{\text{weak}}=52°$ in Appendix I, then after the oblique shockwave: Use Equation (10.16):

$$M_2^2 = \frac{2+(\gamma-1)M_1^2\sin^2\beta}{2\gamma M_1^2\sin^2\beta-(\gamma-1)}\cdot\frac{1}{\sin^2(\beta-\theta)}$$

$$= \frac{2+(1.4-1)\times1.6^2\times(\sin52°)^2}{2\times1.4\times1.6^2\times(\sin52°)^2-0.4}\times\frac{1}{(\sin(52°-10))^2} = 1.45$$

$M_2=1.205$
From Equations (10.20), (10.21), and (10.22):

$$\frac{\rho_2}{\rho_1} = \frac{(\gamma+1)M_1^2\sin^2\beta}{2+(\gamma-1)M_1^2\sin^2\beta} = \frac{(1.4+1)\times1.6^2(\sin52°)^2}{2+(1.4-1)\times1.6^2(\sin52°)^2} = 1.45$$

$\rho_2=1.225\times1.45=1.78$ kgm^{-3};

$$\frac{p_2}{p_1} = \frac{2\gamma}{(\gamma+1)}M_1^2\sin^2\beta - \frac{\gamma-1}{\gamma+1}$$

$$= \frac{2\times1.4\times1.6^2\times(\sin52°)^2}{1.4+1} - \frac{1.4-1}{1.4+1} = 1.69$$

$p_2=1.013\times10^5\times1.69=1.71\times10^5$ Pa;

$$\frac{T_2}{T_1} = 1+\frac{2(\gamma-1)}{(\gamma+1)^2}\frac{(\gamma M_1^2\sin^2\beta+1)}{M_1^2\sin^2\beta}(M_1^2\sin^2\beta-1)$$

$$= 1+\frac{2\times(1.4-1)}{(1.4+1)^2}\times\frac{(1.4\times1.6^2\times(\sin52°)^2+1)}{1.6^2(\sin52°)^2}\times(1.6^2(\sin52°)^2-1)$$

or use the Ideal Gas Law:

$$\frac{p_2}{\rho_2} = R_m T_2, \quad \text{and} \quad T_2 = \frac{p_2}{R_m \rho_2} = \frac{1.71 \times 10^5}{287 \times 1.78} = 334.7 \text{ K};$$

So, $v_2 = M_2 \sqrt{\gamma R_m T_2} = 1.205 \times \sqrt{1.4 \times 287 \times 334.7} = 441.9 \text{ ms}^{-1}$.

d) From Appendix I:
$M_1 = 1.8$, while deflection angle is unchanged, $\theta = 10°$: $\beta_{weak} = 44.5°$.
e) $M_1 = 1.8$, $M_2 = 1$, on the critical Mach line: $\theta = 18.5°$: $\beta = 61.2°$.

Expansion Waves

A compressible airflow expands when it encounters a pressure decreasing disturbance. A pressure expansion wave can be formed. For example, a supersonic airflow travels over a surface, which is deflected away from the direction of the incoming airflow. The deflected surface forms a convex corner with a deflection angle of θ, as shown in Figure 10.7. When a supersonic flow is deflected away from the relative airflow by a very small angle $\delta\theta$, the $\delta\theta$ at the convex corner causes a pressure change, which is a pressure wave, a Mach wave in supersonic flow. When air passes this wave, air pressure decreases, and the airspeed and Mach number of the airflow will increase. This Mach wave is an *expansion wave*. The angle formed between this wave and the airflow is the Mach angle $\mu = \sin^{-1}\left(\frac{1}{M}\right)$, where M is the Mach number of the local airflow. The deflection angle, θ, is the sum of a series (infinite

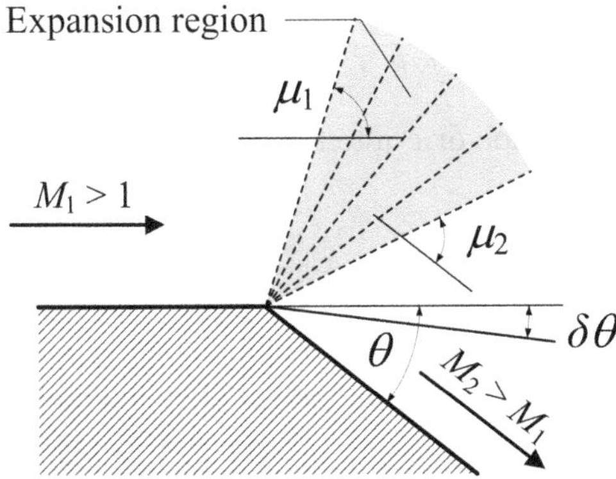

FIGURE 10.7
Expansion waves over a deflected convex corner.

number) of $\delta\theta$s, so a series of corresponding expansion waves will be formed at this convex corner. The Mach number of the local airflow increases after each expansion wave, and the corresponding Mach angle decreases around the corner, which means that each expansion wave propagates away from the previous one. So the series of expansion waves form an *expansion region* around the deflected convex corner. This is also called an *expansion fan*.

Air Properties before and after Expansion Waves

Within an expansion region as shown in Figure 10.7, local airspeed and Mach number gradually increase, and air pressure gradually decreases. The first Mach wave of this expansion region is determined by incoming Mach number M_1, and the Mach angle is angle $\mu_1 = \sin^{-1}\left(\frac{1}{M_1}\right)$; the last Mach wave of the expansion region is determined by the outgoing Mach number M_2, and the Mach angle $\mu_2 = \sin^{-1}\left(\frac{1}{M_2}\right)$. The incoming Mach number M_1 is smaller than the outgoing Mach number M_2, $M_1 < M_2$, and $\mu_1 > \mu_2$.

In the expansion process, air pressure decreases and the air speed increases gradually. There is no heat being transferred to and from the airflow and there is no friction in the expansion process, so this expansion process is an *isentropic process*.

The Energy equation can be used to show the relationship of air temperature and air speed before (denoted as 1) and after (denoted as 2) the expansion region:

$$C_pT_1 + \frac{1}{2}v_1^2 = C_pT_2 + \frac{1}{2}v_2^2 \Rightarrow \frac{\gamma R_m}{\gamma - 1}T_1 + \frac{1}{2}v_1^2 = \frac{\gamma R_m}{\gamma - 1}T_2 + \frac{1}{2}v_2^2;$$

As we know, the speed of sound is a function of temperature of an ideal gas:

$$a = \sqrt{\gamma R_m T} \text{ or } a^2 = \gamma R_m T.$$

Substituting the expression of a^2 into this equation, we can obtain:

$$\frac{T_2}{T_1} = \frac{1 + \frac{\gamma - 1}{2}M_1^2}{1 + \frac{\gamma - 1}{2}M_2^2}, \tag{10.29}$$

which shows the ratio of the temperature after and before the expansion region.

The expansion is a reversible adiabatic/isentropic process, so apply the gas state equations for an adiabatic process to obtain the expressions of air pressure and density ratios before and after the expansion region:

$$\frac{p_2}{p_1} = \left(\frac{1 + \frac{\gamma - 1}{2} M_1^2}{1 + \frac{\gamma - 1}{2} M_2^2} \right)^{\frac{\gamma}{\gamma - 1}} \tag{10.30}$$

and

$$\frac{\rho_2}{\rho_1} = \left(\frac{1 + \frac{\gamma - 1}{2} M_1^2}{1 + \frac{\gamma - 1}{2} M_2^2} \right)^{\frac{1}{\gamma - 1}}. \tag{10.31}$$

Example 10.3

A supersonic air flow at $M_1 = 1.3$ passes a deflected surface and expands, and its Mach number increases to $M_2 = 1.8$ after the expansion region. Assume that the air is under sea level conditions before the expansion region. Find the air temperature, T_2, pressure p_2, and density ρ_2 after it expands.

SOLUTION

Use Equation (10.29) to calculate $T_2 : \frac{T_2}{T_1} = \frac{1 + \frac{\gamma - 1}{2} M_1^2}{1 + \frac{\gamma - 1}{2} M_2^2} = \frac{1 + \frac{1.4 - 1}{2} \times 1.3^2}{1 + \frac{1.4 - 1}{2} \times 1.8^2}$

$= 0.81,$

$T_2 = 0.81 \times 288 = 233.3$ K.

Use Equation (10.30) to calculate $p_2 : \frac{p_2}{p_1} = \left(\frac{1 + \frac{\gamma - 1}{2} M_1^2}{1 + \frac{\gamma - 1}{2} M_2^2} \right)^{\frac{\gamma}{\gamma - 1}}$

$$= \left(\frac{1 + \frac{1.4 - 1}{2} \times 1.3^2}{1 + \frac{1.4 - 1}{2} \times 1.8^2} \right)^{3.5} = 0.48,$$

$p_2 = 0.48 \times 1.013 \times 10^5 = 0.486 \times 10^5$ Pa.

Use Equation (10.31) to calculate $p_2 : \frac{\rho_2}{\rho_1} = \left(\frac{1 + \frac{\gamma - 1}{2} M_1^2}{1 + \frac{\gamma - 1}{2} M_2^2} \right)^{\frac{1}{\gamma - 1}}$

$$= \left(\frac{1 + \frac{1.4 - 1}{2} \times 1.3^2}{1 + \frac{1.4 - 1}{2} \times 1.8^2} \right)^{2.5} = 0.59,$$

$\rho_2 = 0.59 \times 1.225 = 0.723$ kgm^{-3}.

Size of Expansion Region

The size of an expansion region should be related to the deflection angle. Following the explanation of expansion waves, we can see that the greater the deflection angle θ of the surface is, the lower the pressure, and the higher the airspeed, or Mach number, is going to be after the expansion region. German mathematician and engineer, Lugwig Prandtl and Theodor Meyer established the theory of compressible expansion flow, and derived a function known as the *Prandtl–Meyer function,* which describes the relationship between the Mach number of the expanding airflow and the range of turn/angle, the airflow endures in the expansion process:

$$\begin{aligned} \upsilon(M) &= \int \frac{\sqrt{M^2-1}}{1+\frac{\gamma-1}{2}M^2}\frac{dM}{M} \\ &= \sqrt{\frac{\gamma+1}{\gamma-1}}\tan^{-1}\sqrt{\frac{\gamma-1}{\gamma+1}(M^2-1)} - \tan^{-1}\sqrt{M^2-1} \end{aligned} \tag{10.32}$$

where $\upsilon(M)$ is an angle, a function of Mach number M. This angle is also called as *Prandtl–Meyer angle.* The Prandtl–Meyer angles before and after the expansion region can give the value of the deflection angle θ:

$$\theta = \upsilon(M_2) - \upsilon(M_1) \tag{10.33}$$

Appendix III shows the Prandtl–Meyer function as a curve of υ vs M.

Example 10.4

Find the deflection angle of the supersonic airflow described in *example 10.3,* and draw the expansion region.

SOLUTION

Airflow's Mach number before the expansion region is $M_1 = 1.3$, and $M_2 = 1.8$ after the expansion region. Use the Prandtl–Meyer function (10.32):

$$\begin{aligned} \upsilon(M_1) &= \sqrt{\frac{\gamma+1}{\gamma-1}}\tan^{-1}\sqrt{\frac{\gamma-1}{\gamma+1}(M_1^2-1)} - \tan^{-1}\sqrt{M_1^2-1} \\ &= \sqrt{\frac{1.4+1}{1.4-1}}\tan^{-1}\sqrt{\frac{1.4-1}{1.4+1}(1.3^2-1)} - \tan^{-1}\sqrt{1.3^2-1} = 9.85^\circ \end{aligned}$$

$$\begin{aligned} \upsilon(M_2) &= \sqrt{\frac{\gamma+1}{\gamma-1}}\tan^{-1}\sqrt{\frac{\gamma-1}{\gamma+1}(M_2^2-1)} - \tan^{-1}\sqrt{M_2^2-1} \\ &= \sqrt{\frac{1.4+1}{1.4-1}}\tan^{-1}\sqrt{\frac{1.4-1}{1.4+1}(1.8^2-1)} - \tan^{-1}\sqrt{1.8^2-1} = 20.725^\circ \end{aligned}$$

Use Equation (10.32):

$$\theta = \upsilon(M_2) - \upsilon(M_1) = 20.725° - 9.85° = 10.88°$$

The airflow expands and deflects away by 10.88°.

The Mach angle of the first Mach wave: $\mu_1 = \sin^{-1}\left(\frac{1}{M_1}\right) = \sin^{-1}\left(\frac{1}{1.3}\right)$ $= 50.3°$;

The Mach angle of the last Mach wave: $\mu_2 = \sin^{-1}\left(\frac{1}{M_2}\right) = \sin^{-1}\left(\frac{1}{1.8}\right) = 33.7°$;

The expansion region:

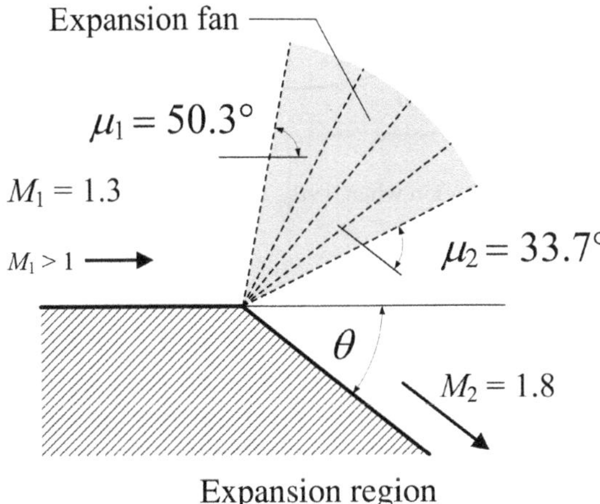

Because $\upsilon(M=1)=0$, $\upsilon(M)$ can be taken as the deflection angle, when air expands increased from Mach number = 1 to Mach number = M.

The Limit of Expansion

Theoretically, when the air expands so much that the pressure decreases to "0", then the air speed will reach the highest possible value, and the expansion region will be at its limit.

Analyze the extreme state with Equation (10.30): p_2 goes to "0" after the expansion region, which means that the Mach number M_2 goes to "∞", infinity. So the maximum deflection angle associates with the expansion that airflow at $M_1 = 1$expands to $p_2 = 0$ Pa, and $M_2 \to \infty$. From Equation (10.33), we know that $\upsilon(1) = 0°$, and:

$$\theta_{max} = \lim_{M_2 \to \infty}\left(\sqrt{\frac{\gamma+1}{\gamma-1}}\tan^{-1}\sqrt{\frac{\gamma-1}{\gamma+1}(M_2^2-1)} - \tan^{-1}\sqrt{M_2^2-1} - 0\right)$$

$$= \sqrt{\frac{\gamma+1}{\gamma-1}} \times \frac{\pi}{2} - \frac{\pi}{2} = \frac{\pi}{2}\left(\sqrt{\frac{\gamma+1}{\gamma-1}} - 1\right)$$

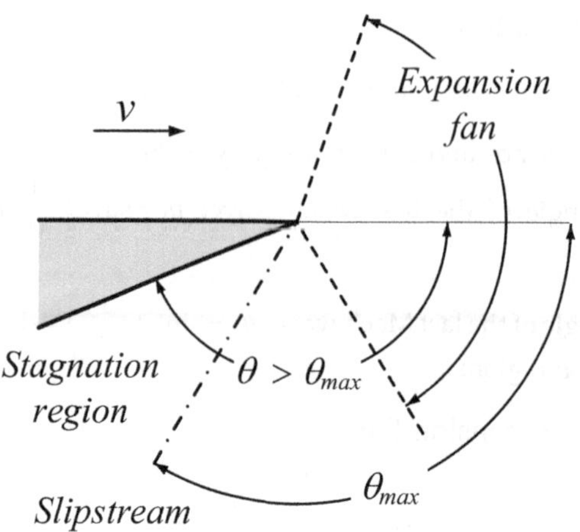

FIGURE 10.8
Slipstream and the stagnation region when $\theta > \theta_{max}$.

For air $\gamma = 1.4$, $\theta_{max} = 1.45\frac{\pi}{2} = 130.45°$. This is the extreme limit of deflection, and the air cannot expand further even if $\theta > \theta_{max}$. Figure 10.8 shows the situation where $\theta > \theta_{max}$. Air cannot not expand beyond $\theta = \theta_{max}$. The line at the end of θ_{max} is called the *slipstream,* and the area between the end of the deflection angle θ and the slipstream is called the *stagnation region,* as shown in Figure 10.8. In theory, the pressure p at the slipstream is "0", so as the pressure in the stagnation region; the speed of the airflow at the slipstream reaches its maximum value, however the air flow speed in the stagnation region is "0". In a real airflow, such a sudden change in speed cannot be sustained due to the viscous nature of air. Therefore, in practice, there would be vortices formed in the stagnation region. Those vortices could cause oscillation locally and corrosion of the surface due to fluctuating pressure and temperature in the stagnation region.

Example 10.5

There is a supersonic airflow, which passes a convex-deflected corner as shown in Figure 10.7. The incoming Mach number is $M_1 = 1.5$ under sea level conditions, the deflection angle is $\theta = 15°$. Find the Mach number after the expansion fan M_2, and the air property p_2, ρ_2, and T_2, and the air speed after the expansion fan.

SOLUTION

Equation (10.33) $\theta = \upsilon(M_2) - \upsilon(M_1)$ gives the relationship between the Mach numbers M_1, M_2, and the deflection angle θ. However, carrying out the inverse calculation of the Prandtl–Meyer function (10.32) cannot be easily done by a calculator. Students can use the table showing the

Prandtl–Meyer function in Appendix II to assist the calculation required in this example.
Find $\upsilon(1.5) = 11.9°$ in Appendix II, so $\upsilon(M_2) = 15° + 11.9° = 26.9°$. Go back to Appendix II, column υ, to find a nearest number to 26.9, is 26.9295. The corresponding Mach number is 2.02.
So $M_2 = 2.02$.
Air property is under sea level conditions before the expansion: $p_1 = 1.013 \times 10^5$ Pa, $\rho_1 = 1.225 \text{kgm}^{-3}$, and $T_1 = 288$ K.

Use Equation (10.29) to calculate T_2: $\dfrac{T_2}{T_1} = \dfrac{1+\dfrac{\gamma-1}{2}M_1^2}{1+\dfrac{\gamma-1}{2}M_2^2} = \dfrac{1+\dfrac{1.4-1}{2}\times 1.5^2}{1+\dfrac{1.4-1}{2}\times 2.02^2}$

$= 0.798,$

$T_2 = 0.798 \times 288 = 229.9$ K.

Use Equation (10.30) to calculate p_2: $\dfrac{p_2}{p_1} = \left(\dfrac{1+\dfrac{\gamma-1}{2}M_1^2}{1+\dfrac{\gamma-1}{2}M_2^2}\right)^{\frac{\gamma}{\gamma-1}}$

$$= \left(\dfrac{1+\dfrac{1.4-1}{2}\times 1.5^2}{1+\dfrac{1.4-1}{2}\times 2.02^2}\right)^{3.5} = 0.455,$$

$p_2 = 0.455 \times 1.013 \times 10^5 = 0.46 \times 10^5$ Pa.

Use the Ideal Gas Law to calculate p_2: $\dfrac{p_2}{\rho_2} = R_m T_2$, then $\dfrac{0.455\times 10^5}{\rho_2}$

$= 287 \times 229.9,$

$\rho_2 = 0.69 \text{ kgm}^{-3}$.
The air speed after the expansion region υ_2:

$$\upsilon_2 = M_2 a2 = M_2\sqrt{\gamma R_m T_2} = 2.02\sqrt{1.4\times 287\times 229.9} = 613.9 \text{ ms}^{-1}$$

Exercises

1. When a supersonic airflow, $M = 1.8$, passes through a normal shockwave under sea level conditions, what are the values of the stagnation pressure before and after the normal shockwave?
2. The Mach number of an airflow is $M_1 = 1.4$, air pressure $p = 1.012 \times 10^5$ Pa, and air temperature is 273 K. This airflow, then, passes through an oblique shockwave on a deflected surface, and the Mach number of the airflow decreases to $M_2 = 1$.
   (a) Find the deflection angle. Is the oblique shockwave attached to, or detached from the deflected surface?

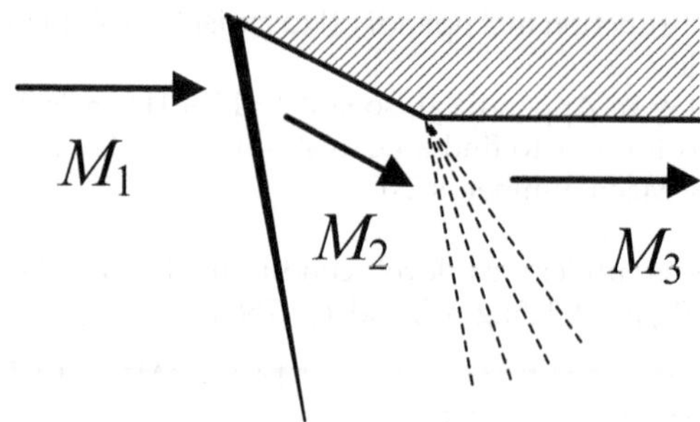

FIGURE 10.9
Supersonic airflow over a twice-deflected surface.

(b) Find the shock angle;

(c) Calculate the air pressure, temperature, density, and air speed after the shockwave.

3. The Mach number of an airflow is $M_1 = 1.4$, air pressure $p = 1.012 \times 10^5$ Pa, and air temperature is 273 K before an expansion region. The deflection angle of the expansion region is, $\theta = 20°$.

   (a) What is the Mach number of the airflow after the expansion?

   (b) Sketch a diagram to show the expansion fan, which starts from the first Mach line, and end with the last Mach line.

   (c) Calculate the air temperature, pressure, and density, and airspeed after the expansion fan.

4. The airflow passes a twice-deflected surface. The free-stream Mach number, M_1 is 2 at sea level, and the air flows along the path with a 16° deflection angle passes through a weak oblique shockwave; and then passes a convex corner back, where the surface is deflected back to horizontal, as shown in Figure 10.9.

   a) What is v_1?

   b) Find the final Mach number M_3.

   c) Find the air speed v_3.

11

Introduction of Supersonic Flight

The wings of a supersonic airplane are very different from those of subsonic airplanes or transonic airplanes. The difference is in their appearance, and more importantly in the principle of generation of lift. From previous chapters, we have learned that a bow shockwave will be formed at the leading edge of an aerofoil when the free-stream Mach number M_{fs} reaches the detachment Mach number M_{det} for transonic aircraft. Lift would be reduced and drag increase significantly, if the free-stream Mach number was to increase further. As a result, an aircraft could not sustain the usual performance using an aerofoil designed for subsonic or transonic flight. Obviously, a different principle of lift generation is required for any aircraft, which can fly with a Mach number greater than the detachment Mach number, and shape of the wing needs to be altered accordingly.

Two types of waves, oblique shockwave and expansion wave, can be formed in a supersonic airflow, as discussed in Chapter 10. The air pressure increases when the airflow passes oblique shockwaves, and decreases when it passes expansion waves. Lift can be generated if oblique shockwaves and expansion waves are formed around an aerofoil at assigned locations by design. This chapter is only an introduction to supersonic flight as an application of the knowledge gained from Chapter 10.

Supersonic Flow over Aerofoil

Oblique shockwaves or expansion waves can be formed on the surface of an aerofoil, as long as the surface deflects to, or from, the supersonic airflow. There is nearly no limit to the shape of supersonic aerofoil apart from it should be slim. However, the shape of aerofoil will affect the pattern of formation of the waves, but it is not the determining factor on whether the aerofoil is able to produce lift.

Patterns of formation of oblique shockwaves and expansion waves for several possible aerofoil designs are described and analyzed in this chapter.

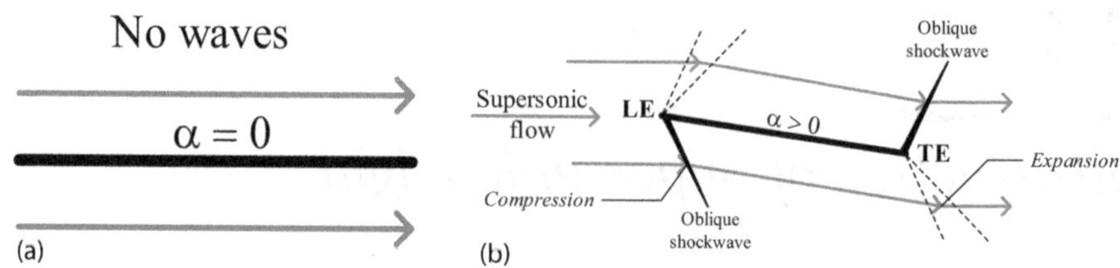

FIGURE 11.1
A plate in a supersonic airflow: (a) angle of attack $\alpha = 0$; (b) at a small positive angle of attack.

Thin Plate

A plate lays in parallel to the airflow, i.e. the angle of attack is "0", as shown in Figure 11.1 (a). The airflow around the plate is hardly disturbed, and no waves are formed on the plate. There is no change in air pressure around this plate. So there is no pressure difference generated on both sides of this plate – no lift.

When this thin plate is in a supersonic airflow with a small positive angle of attack as shown in Figure 11.1 (b), the upper surface of the plate is deflected away from the free-stream airflow, and an expansion fan is formed on the upper end of leading edge. The first Mach line of the expansion fan is determined by the Mach number of the incoming airflow, and the last Mach line will be determined by the Prandtl–Meyer function and Equation (10.33). The airspeed increases and the air pressure decreases when airflow passes through the expansion fan. The lower surface of the plate is deflected toward the free-stream airflow, so an oblique shockwave is formed on the lower end of leading edge. The air pressure increases after the oblique shockwave.

Similarly, at the trailing edge, an oblique shockwave is formed on the upper end of trailing edge, while an expansion wave fan is formed on the lower end of trailing edge.

It is very clear that air pressure over the upper surface of this plate behind the expansion fan in Figure 11.1 (b) is lower than the air pressure behind the oblique shockwave on the lower surface. This pressure difference between the two sides of the plate will to produce lift in a supersonic airflow.

At the trailing edge of the plate in Figure 11.1 (b), air pressure increases after the compressive oblique shockwave on the top end, and air pressure decreases after the expansion waves on the lower end. Those changes in air pressure at the trailing edge would not affect the lift production of the plate. However, they do contribute to the total drag of the plate in the airflow.

Pressure Coefficient

Pressure coefficient is defined as $C_p = \frac{p - p_{fs}}{q}$, where p is local pressure; p_{fs} is the air pressure in free-stream; and q is the dynamic pressure. In general, the distribution of pressure coefficient represents the pressure distribution around an aerofoil. The characteristics of the distribution of pressure coefficient show the capability of the aerofoil to produce lift.

The supersonic linearized theory can explain the change in air pressure around a supersonic aerofoil in a relatively simple equation. This theory defines a supersonic aerofoil: there is no shockwaves on the surface of a thin sharp-edged aerofoil with small camber. The pressure coefficient for this aerofoil follows *Ackeret's rule: at low incidence in 2-D frictionless shock-free supersonic flow, the pressure coefficient is:*

$$C_p = \frac{2\varepsilon}{\sqrt{M^2 - 1}} \tag{11.1}$$

where M is the Mach number of the flow; ε is the angle (in radians) between the local airflow and the local tangent of the surface, which illustrated in Figure 11.2.

Angle ε is negative "–" if the surface tangent deflects away from the incoming airflow; Angle ε is positive "+" if the surface tangent deflects toward the incoming airflow.

For example, if we analyze the pressure distribution around the thin plate shown in Figure 11.1 (b), ε is a constant negative value on the upper surface of the plate, and ε is a constant positive value on the lower surface. The direction of pressure difference is always heading from the lower surface C_p to upper surface C_p. Its distribution of pressure coefficient is shown in Figure 11.3, which indicates this plate generates lift.

There are more sophisticated methods to analyze the lift generated by a supersonic aerofoil with the full knowledge of the locations and features of the supersonic waves formed around the supersonic aerofoil. The supersonic airflow around a supersonic aerofoil can be relatively accurately described by an aerodynamic model – a set of mathematical equations and their boundary and initial conditions. The model can be solved by well-developed numerical simulation methods to obtain the velocity and air properties at any point and at any time of the airflow field. However, the scope of this chapter is to try to explain the lift principle of a supersonic flight in a plain and simple mathematical expression. Readers need to take an advanced aerodynamics courses if more detailed aerodynamic knowledge for supersonic flight is required.

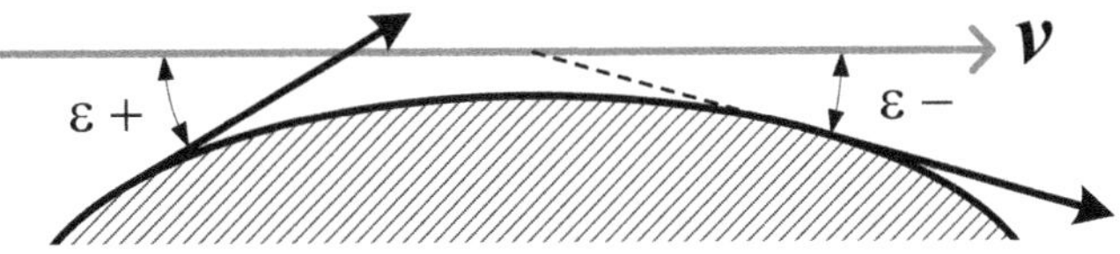

FIGURE 11.2
Illustration of positive and negative ε.

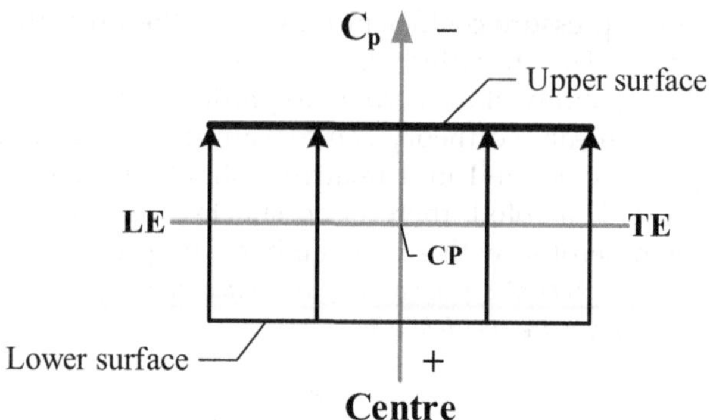

FIGURE 11.3
Distribution of pressure coefficient on a plate with a positive angle of attack.

Double Wedge

Figure 11.4 (a) shows a symmetrical aerofoil with a double wedged cross section sitting in a supersonic airflow, and the angle of attack of this aerofoil to the free-stream airflow is "0". The upper surface and the lower surface are symmetrical, and the pattern of the compressive oblique shockwaves and expansion waves formed on both surfaces are symmetrical: air pressure increases from the leading edge after the compressive oblique shockwave; then it decreases when air passes through the expansion region. The pressure distributions over the upper surface and the lower surface of the double wedge are identical, so there is no lift generated.

The wedge angle at leading edge and trailing edge is relatively small. In Figure 11.4 (b), the double wedge aerofoil sets in the supersonic airflow with a small angle of attack, e.g. a half wedge angle, and the number 1, 2, 3, and 4 represents these four flat parts of the surface. The upper surface "1" is in parallel to the incoming airflow, and the air flows through without any disturbance, so there is no supersonic wave formed at the upper end of leading

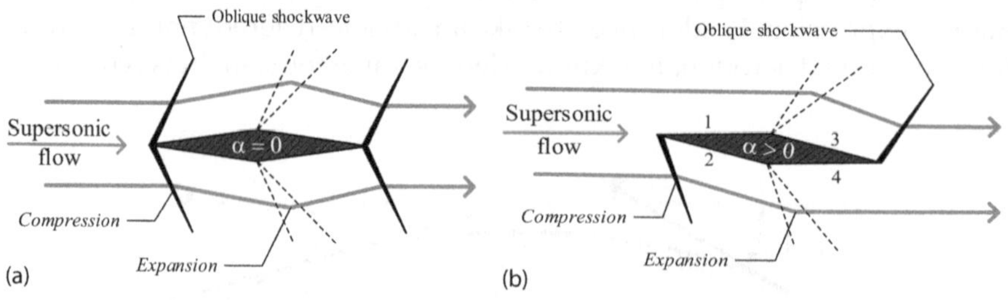

FIGURE 11.4
Double wedge aerofoil in a supersonic airflow: (a) angle of attack $\alpha = 0$; (b) at a small positive angle of attack.

edge, and the pressure over surface "1" is the same as that in free-stream: $p_1 = p_{fs}$. At the lower end of leading edge, airflow is deflected toward the airflow – an oblique shockwave formed, and the air pressure after the oblique shockwave increases, so the air pressure over the surface "2" $p_2 > p_{fs}$.

At the middle upper wedge between surface "1" and "3" of the aerofoil, and the middle lower wedge between surface "2" and "4", expansion fans are formed, because both surface "3" and "4" deflect away from the airflow. Surface "4" is parallel to the free-stream airflow again. At the upper end of trailing edge, an oblique shockwave is formed. The pressure over the upper surface "3" is lower than the free-stream pressure: $p_3 < p_{fs}$, and the pressure over the surface "4" is lower than the pressure over surface "2", but it is still little greater than the free-stream pressure: $p_4 \geq p_{fs}$ (see Exercise 4 in Chapter 10).

Summarizing the air pressure over the upper surface and lower surface of the double wedge aerofoil: $p_1 = p_{fs} < p_2$, and $p_3 < p_{fs} \leq p_4$. So, the overall pressure distribution produces a total force upward – lift.

Applying Ackeret's rule, the distribution of pressure coefficient for the double wedge aerofoil in Figure 11.4 (a) is shown in Figure 11.5 (a). The angle ε on both upper surface and lower surface of the front wedge are the same at the one location on the chord, so the distributions of pressure coefficient for both upper and lower surface are the same. As a result, the pressure distribution shown in Figure 11.5 (a) does not produce lift.

For the double wedge aerofoil in Figure 11.4 (b), the angle of attack of the aerofoil to the relative airflow is a small, assuming the top leading edge is parallel to the airflow. According to the Ackeret's rule, ε on upper surface "1" is "0"; ε on upper surface "3" is a "–" constant; ε on lower surface "2" is a "+" constant; and ε on lower surface "4" is "0". The distribution of pressure coefficient around the aerofoil is in Figure 11.5 (b), where the arrows indicate the direction of the pressure difference between the lower and upper surface. It is clear that the double wedge aerofoil with a small positive AoA in a supersonic airflow produces lift. Please note that the center of pressure, CP

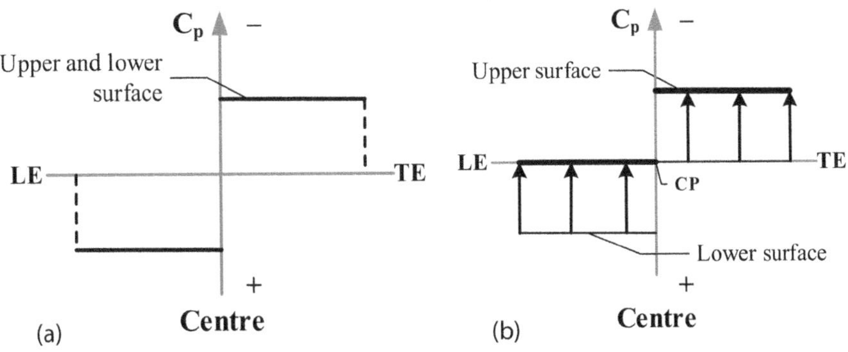

FIGURE 11.5
Distribution of pressure coefficient on a double wedge aerofoil in a supersonic airflow (a) angle of attack $\alpha = 0$; (b) at a small positive angle of attack.

(center of the arrow area), for the symmetrical double wedge aerofoil lies at the middle of the chord. It is at the position of vertical axis in Figure 11.5 (b).

This demonstrates that the double wedge aerofoil can produce lift. Double wedge aerofoil was a popular design for early supersonic flight. The shape of double wedge can be symmetrical or cambered – nonsymmetrical as shown in Figure 11.6, where t is the maximum thickness of the aerofoil, and c is the length of chord. The selected ratio of thickness of t/c to an aerofoil depends on the Mach number of the supersonic flight. For example, the ratio of $t/c=4\%$ is best for the supersonic flight of $M=1.3$, suggested in Houghton, Carpenter et al. (2012).

The location of maximum thickness has almost no effect on lift and CP (center of pressure) for a symmetrical double wedge aerofoil. But the location of maximum thickness/camber does affect the lift coefficient and the center of pressure to a cambered double wedge aerofoil.

Figure 11.7 shows a symmetrical double wedge aerofoil with forward maximum thickness (a), and a cambered double wedge aerofoil with centered maximum thickness (b), and a cambered double wedge aerofoil with forward maximum thickness (c). These three aerofoils sit in a supersonic airflow, $v>a$, at a small angle of attack. The pressure distributions of the aerofoils are shown in Figure 11.8 (a), (b), and (c), respectively.

For a double wedge aerofoil, the location of maximum camber is the same as the location of maximum thickness. It is can be observed that the CP remains at the center of the chord as shown in Figure 11.8 (a), even when it has a forward maximum thickness. On the other hand, Figure 11.8 (b) and (c) shows that the center of pressure moves away from the middle of the chord for cambered aerofoils, and the location of CP changes when the location of the maximum thickness changes.

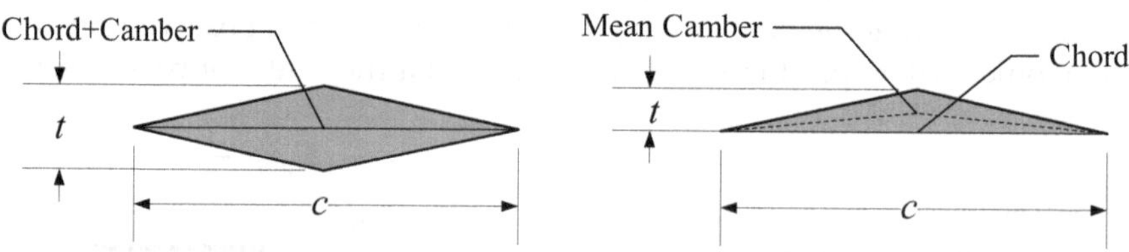

FIGURE 11.6
Symmetrical and cambered double wedge aerofoil.

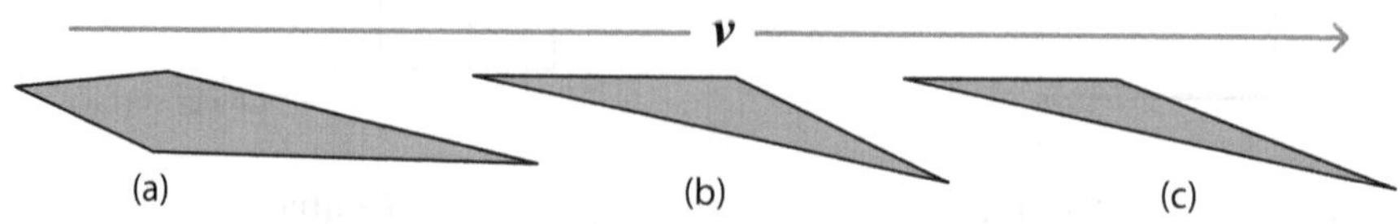

FIGURE 11.7
(a) Symmetrical double wedge aerofoil with forward maximum thickness at a small positive α; (b) Cambered double wedge aerofoil with centered maximum thickness at a small positive α; (c) Cambered double wedge aerofoil with forward maximum thickness at a small positive α.

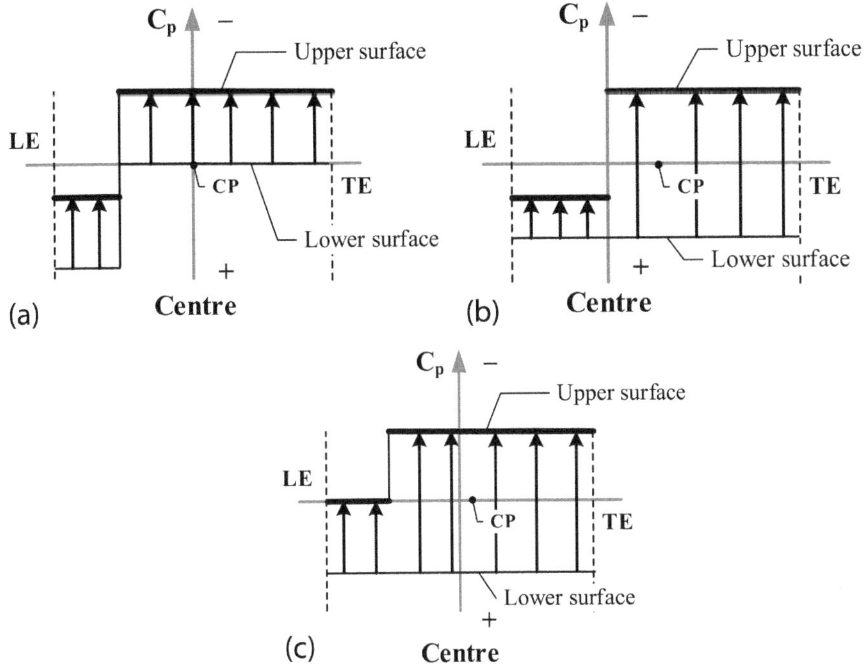

FIGURE 11.8
(a) Distribution of pressure coefficient C_p around the aerofoil in Figure 11.7 (a); (b) Distribution of pressure coefficient C_p around the aerofoil in Figure 11.7 (b); (c) Distribution of pressure coefficient C_p around the aerofoil in Figure 11.7 (c)

Biconvex

Double wedge aerofoil was a popular aerofoil in the early development of supersonic flights, and another popular supersonic aerofoil is biconvex aerofoil, which was commonly used for $M = 1.4 \sim 2$ flights. Two circular arcs form the upper surface and lower surface of the aerofoil. Two arcs with a same curvature (radius) form a symmetrical biconvex aerofoil, as shown in Figure 11.9 (a), while arcs with different radii of curvature form a cambered aerofoil, as shown in Figure 11.9 (b). The leading edge of the biconvex is relatively sharp, and shockwave(s) will be formed and attached at the leading edge.

When a symmetrical biconvex aerofoil sits in a supersonic airflow with a small angle of attack, as shown in Figure 11.10 (a), an oblique shockwave is formed at the lower end of leading edge, and another oblique shockwave is formed at the upper end of trailing edge. Both the upper and lower surfaces are covered with gradually deflected expansion waves. As we have learned in the previous chapters, air pressure will increase after a shockwave, and will decrease after expansion waves. So the air pressure on top of the aerofoil decreases from free-stream pressure, p_{fs}, while the pressure on the

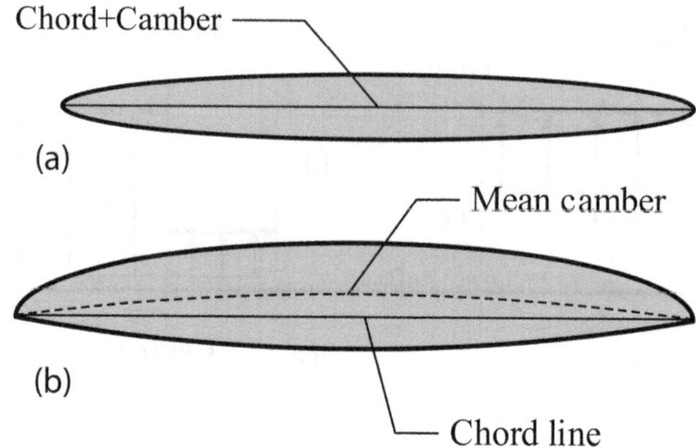

FIGURE 11.9
(a) Symmetrical biconvex aerofoil; (b) Cambered biconvex aerofoil.

lower surface gradually decreases from the pressure after the shockwave, which is much greater than p_{fs}. The pressure coefficient can be determined by Ackeret's rule, Equation (11.1). The pressure coefficient C_p on the upper surface of the biconvex aerofoil is negative, and the C_p on the lower surface is positive. The distribution of the pressure coefficient of this symmetrical biconvex aerofoil is shown in Figure 11.10 (b). The distribution shows that this aerofoil produces lift.

A biconvex aerofoil can be cambered, as seen in Figure 11.9 (b). Usually, the camber is a positive camber. For the same small angle of attack as shown in Figure 11.11 (a), oblique shockwaves will be formed at lower LE and upper TE. The pressure coefficient C_p can be determined by the Ackeret's rule (11.1), and the distribution of pressure coefficient C_p of a cambered biconvex aerofoil is shown in Figure 11.11 (b).

Results from laboratory tests have shown that a biconvex aerofoil can work at a relatively high angle attack, as high as approximately 30°. It has become

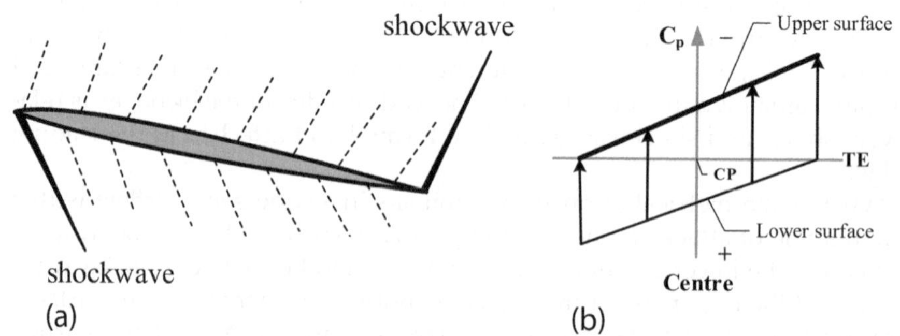

FIGURE 11.10
(a) Symmetrical biconvex aerofoil at small positive α; (b) Distribution of pressure coefficient C_p around the aerofoil in (a).

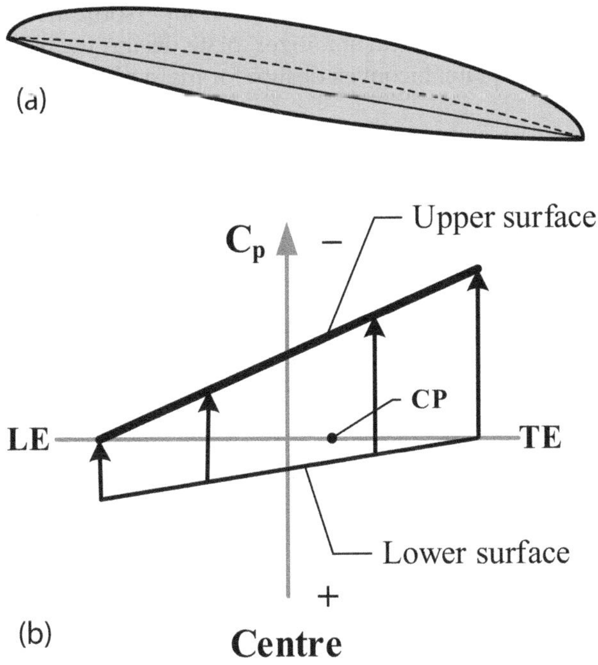

FIGURE 11.11
(a) Cambered biconvex aerofoil at small positive α; (b) Distribution of pressure coefficient C_p around the aerofoil in (a).

widely used in the design of supersonic aerofoil. To analyze the characteristics of aerodynamic forces is relatively easy, because experiments have shown that the aerodynamic parameters obtained from aerofoil test results agree with those derived from aerodynamic theories as stated in Houghton, Carpenter et al. (2012).

If the angle of attack is too small, oblique shockwaves can be formed on both sides of leading edge of biconvex aerofoil. But the shockwave on the leading edge of upper surface is much weaker than the shockwave at the leading edge of lower surface. However, the gradually deflected expansion waves cover both of the upper and lower surfaces.

The shapes of supersonic aerofoil discussed here are popular, and are well used. A plate design of aerofoil is simple and economical, but it has a poor rigidity, which would affect the stability of the flight. Double wedge design is very popular choice, particularly, for early generations of supersonic aircraft. It can be manufactured easily and is strong. Biconvex design is increasingly used in modern supersonic aircraft. The shape of biconvex aerofoil is similar to that of sub/transonic aerofoil, and it should perform better in subsonic speeds than plate or double wedge aerofoil.

Different types of aerofoil can be used in one supersonic aircraft: for example, biconvex design for vertical stabilizer, plate design for horizontal stabilizer, and cambered plate/biconvex design for main plane.

Boundary Layer and Drag

From this analysis and discussion, we realize that there is nearly no limitation on the shape of supersonic aerofoil, for example, hexagonal aerofoil, as long as supersonic waves can be organized in a pattern, which enable to produce aerodynamics force TR (total reaction). Even the body of a supersonic airplane can contribute to the force with its wings due to the shockwaves and expansion waves formed around the entire body.

Boundary Layer in Supersonic Flow

Since air is a viscous fluid, a boundary layer will be formed over the surface of an aircraft when it travels at a supersonic speed. However, the boundary layer over a supersonic flight will be very thin. The speed of air particles on the surface under the boundary layer is "0", and should increase very quickly to reach the free-stream speed, so it can be assumed that the airflow speed increases with the boundary layer linearly in the direction perpendicular to the surface.

Boundary layer separation of a supersonic airflow can occur, particularly the turbulent wake after a shockwave separates immediately after a strong shockwave. As discussed in the section on supersonic airflow, the pressure distribution around the supersonic aerofoil depends primarily on the formation of different supersonic waves. The changes in pressure after those oblique shockwaves and expansion waves produce the lift. The boundary layer separation is a part of the pressure change. Of course, the separation will cause the increase in drag due to the wake of vortices.

The viscous friction in the boundary layer on the surface of a supersonic aircraft, namely skin drag, is a complex issue. The viscous friction per area can be determined by Equation (3.1): $\tau = \mu \frac{\partial u}{\partial y}$, where $\frac{\partial u}{\partial y}$ is the change of local airspeed in the boundary layer, u, in the direction perpendicular to the surface, y. $\frac{\partial u}{\partial y}$ is significant within the turbulent boundary layer, so a part of the kinetic energy of the airflow has to be consumed to overcome the viscous friction – skin drag, and this part of kinetic energy will be dissipated as heat. The airflow within the boundary layer will be heated up, and the temperature of the air within the boundary layer will be increased. As we know from Chapter 3, the dynamic viscosity, μ, of air will increase with the increase of temperature, which will cause the increase of the viscous friction further.

Drag of Supersonic Flight

There are three parts of drag to a supersonic airplane:

$$D = D_{fr} + D_{\text{wave}} + D_{\text{in}}$$

D_{fr}, friction drag caused by the viscous nature of air with boundary layer, mainly considered as skin drag.

D_{wave}, wave drag caused by formation of shockwaves around the whole body of the airplane. The formation of a shockwave, which is part of the makeup of producing lift, consumes energy. This energy is from the kinetic energy of the supersonic airflow. To the aircraft, it is a drag. Wave drag is a significant part of the drag on supersonic flight.

$D_{in} = D_{vortex} + D_{lift}$, "induced" drag is associated with lift, which includes drag. The pressure distribution around supersonic aerofoil/aircraft is formed by the sequence pattern of oblique shockwaves and expansion waves. The vortices in the turbulent wake behind shockwaves causes drag, D_{vortex}. Integration of the pressure distribution over the surface of a supersonic aircraft results in a force, F, shown in Figure 11.12. The F is perpendicular to the aircraft. It can be divided in to two components: one in vertical direction – lift, and the other in horizontal direction (RA) – drag, companying the lift, D_{lift}. The "induced drag" related to the pressure distribution is the sum of D_{lift} and D_{vortex}.

Wave drag D_{wave} is the dominate component of the total drag to a supersonic aircraft. While not dominate, D_{fr} is a source of kinetic heating. C_D of supersonic flight is relatively high, so the ratio of L/D is relatively low compared with that of sub/transonic aerofoils.

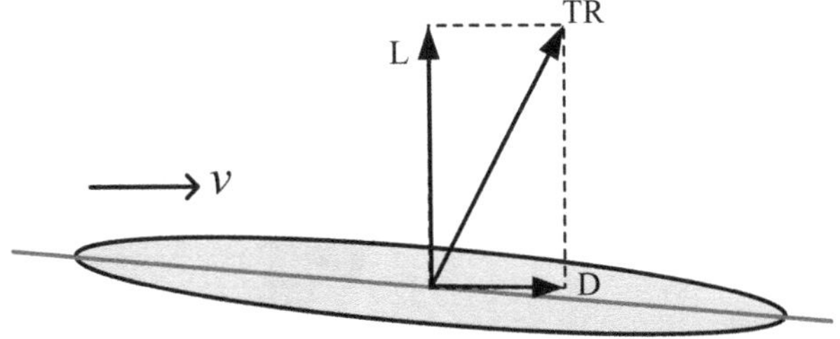

FIGURE 11.12
The forces from the pressure distribution on a supersonic aerofoil.

Supersonic Wings and Planforms

Planforms for supersonic aircraft varies widely. For example, it could be sweep back wings, like Concord; it could be delta wing, which wing and fuselage are integrated into one triangle shape; and there are straight wing and other wing/fuselage combination designs for jet fighters. As an introduction to supersonic aircraft, only a few examples of planform designs will be discussed here.

Unswept Wings

An unswept (straight) straight wing can be used in supersonic aircraft, for example, NASA's FA 18, as shown in Figure 11.13. The straight wing can produce higher lift coefficient at low airspeed range, and relatively lower drag coefficient at a high Mach number in supersonic speed range.

However, a straight wing's performance can be weakened by tip effects. The airflow around a straight wing is approximately two-dimensional (2-D) flow. Its pressure distribution can be analyzed as earlier except at the wing tips. At the wing tip areas of a straight wing, the local airflow is 3-D because the formation of supersonic waves are affected by the tip boundary and vortices in the tip area, and the lift produced in the tip area will be reduced, this is called (supersonic) *tip effect*. Furthermore, the wing tips can easily end up situated in the silent zone outside the Mach cone of the aircraft when the Mach number increases if the wingspan is relatively long with the consequence that lift will be reduced and its drag increased. So the design of straight wing may not be suitable for high supersonic aircraft.

FIGURE 11.13
Image of NASA's FA-18 with straight wings (Credit: NASA).

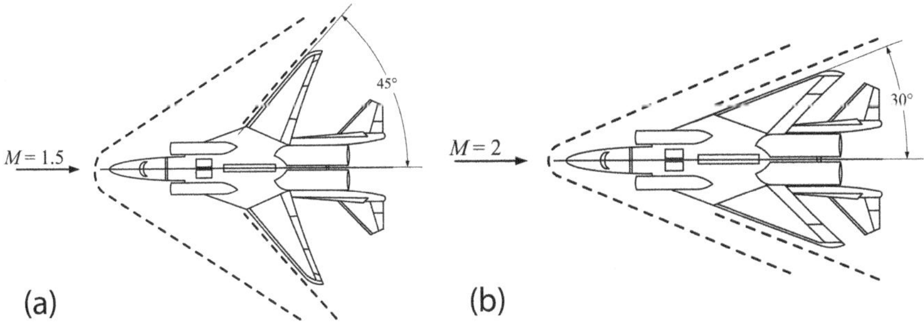

FIGURE 11.14
Supersonic aircraft with swept-back wings in Mach cone (a) $M_{fs}=1.5$ and $\mu<\sigma$; (b) $M_{fs}=2$, and $\mu=\sigma$.

Swept Wings

It is the common design for supersonic aircraft. When an aircraft travels at a supersonic speed, a Mach cone will be formed from the nose of a supersonic aircraft, as described in Chapter 6. Swept wings are better situated inside the Mach cone as shown in Figure 11.14, in which μ is Mach angle, Λ is sweep angle, and angle $\sigma=90° - \Lambda$. In Figure 11.14 (a), M_{fs} is 1.5, $\mu=41.8°$, and the Λ is 45°, so $\mu<\sigma$. In Figure 11.14 (b), M_{fs} is 2.0, $\mu=30°$, and the Λ is 60°, so $\mu=\sigma$. The wings would be heavily swept back if the aircraft travel at a higher Mach number. When $\mu\geq\sigma$, as shown in Figure 11.14 (b), the leading edge of the wing is likely in sonic or subsonic flow (chordwise). Its drag will be relatively low, and produce high ratio of *L/D*. However, the rigidity of the heavily swept-back wings will be low, and the wings would experience twist and bending problems.

When $\mu<\sigma$, as shown in Figure 11.14 (a), the leading edge of the swept wing would be likely in supersonic flow (chordwise). Oblique shockwaves would be formed at the leading edge close to the wing root. After an oblique shockwave airflow is still supersonic at the leading edge of the wing, so a series of shockwaves can be formed along the leading edge of the swept wing as shown in Figure 11.15. Therefore, the shock drag of a swept wing would increase considerably at a high Mach number. The tip effect in a supersonic leading edge on a swept wing would reduce lift and make a low *L/D* ratio at a high Mach number.

Delta Wings

A delta wing is a design to join the fuselage and wings together forming a triangle. Many modern supersonic jets have chosen delta wing design. Examples of delta wings are shown in Figure 11.16. This design has the aerodynamic character of swept wings, but it can make the supersonic wing rigid. The structure of a delta wing is simple, and mechanically

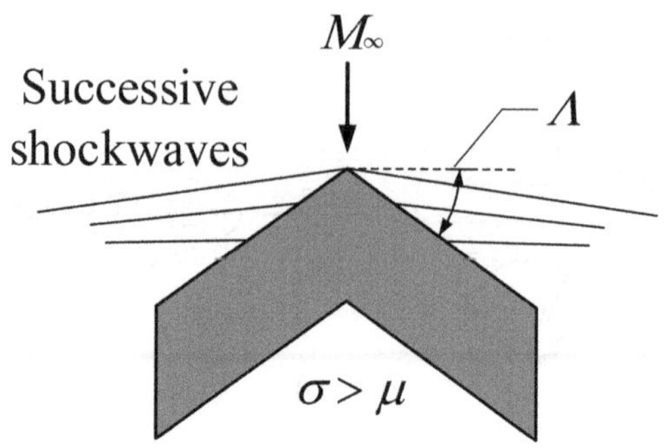

FIGURE 11.15
Multiple shockwaves on LE of swept wing when $\mu < \sigma$.

FIGURE 11.16
(a) Delta wing design; (b) Image of Tu-144LL, Russian supersonic jet with delta wing design (Credit: NASA); (c) New concept of future supersonic passenger jet with delta wing (Credit: NASA/Lockheed Martin).

strong. The chord is relatively long, and the relative thickness *t/c* is low. The area of a delta wing is large, so the wing load is relatively low. However, a "large area" will produce higher skin friction – higher skin drag.

Variable Sweep Wings

Some supersonic aircraft are designed to be able to cruise in both subsonic and supersonic speed range. Straight wing could produce better L/D ratio in subsonic range, while swept wings would perform better in supersonic range as discussed above. Variable sweep wings, or swing wings were developed to ensure aircraft have sufficient capability both in sub- and supersonic speeds. The concept of variable sweep wings is shown in Figure 11.17. A variable geometry wing can change its configuration in flight: the wing swings up to become a straight wing at a subsonic airspeed, and folds back toward to its fuselage to become a sweep back wing in a supersonic airspeed.

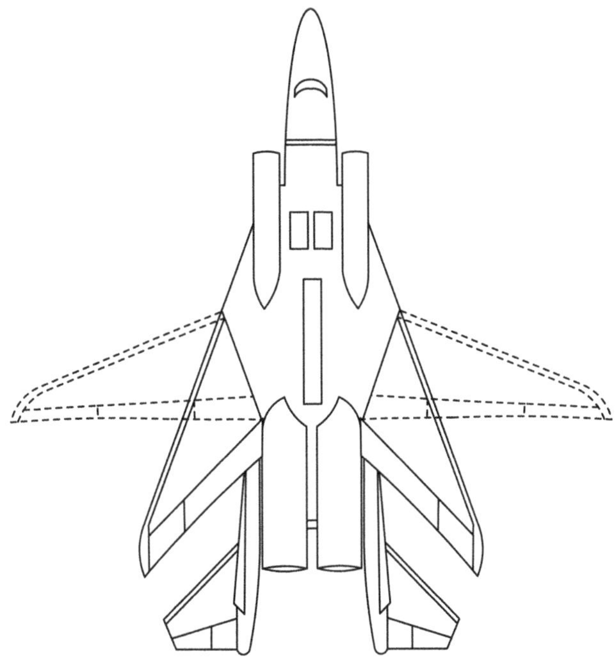

FIGURE 11.17
Variable sweep wing design.

Body Shapes of Supersonic Airplanes

The design of a body of a supersonic airplane should follow the area rule to reduce drag. A slender shape with relative sharp leading edge/nose and trailing edge is always preferred design, as indicated by dashed lines in Figure 11.18. Sharp leading edge can make bow shockwave formed at leading edge attached to the surface, which would cause less drag than the detached bow shockwave from the leading edge does if the leading edge is round.

However, a leading edge is a source of kinetic heating, and the thermal load and intensity would be very high at a sharp leading edge. The actual body of a supersonic aircraft would have the body shape similar to the solid line profile shown in Figure 11.18. For the actual body of a supersonic aircraft, the leading edge is not sharp but instead rounded with a small radius, and

FIGURE 11.18
Body shape of supersonic aircraft following area rule.

jet engine situates at tailing edge with a cut off trailing edge as jet exhaust. Designing the body of a supersonic plane needs to consider the aspects of aerodynamic forces and heating comprehensively.

Kinetic Heating

Heating on the surface of a supersonic aircraft is a special issue, which subsonic and transonic aircraft do not have. One of the examples is the high temperature a space shuttle experienced when it came back from space and reentered the Earth atmosphere at a Mach number > 20. This heating is called *kinetic heating*. Namely, kinetic heating is the process that kinetic energy turns into thermal energy – heat, or that the decrease of supersonic airspeed results in the increase of air temperature. The sources of kinetic heating on supersonic flights are:

1) Friction with boundary layer: the kinetic energy of the airflow would be consumed to overcome the friction for air to keep flow forward. The work done to against friction will change into thermal energy – heat. The surface the friction exerted on will be heated, and it temperature will be increased.

   Formula (11.2), suggested by Kermode (2012), shows that the temperature increase is proportional to the square of airspeed in kt:

   $$\Delta T = (v/100)^2 (^\circ\mathrm{C}) \tag{11.2}$$

   For example: an aircraft travels at sea level condition, and its speed increases to $M=2$, i.e. $v=1323$ kt. The temperature on its surface will increase by 175°C, which can make some plastic components soft and change in shape. If the Mach number is over 5, the surface temperature may increase by more than 1000°C.

2) Stagnation points: According to the energy equation: $C_pT + \frac{1}{2}v^2 = E$, at a stagnation point, $v=0$ ms^{-1}, temperature will increase.

   For example, at the leading edge of an aircraft which travels at $M=2$ (sea level condition). Its temperature at leading edge can be as high as 520 K (close to 250°C). So for shuttle reentry, the leading edge of the shuttle can experience temperature of approximate 20,000 degrees, extremely high temperature.

3) Shockwave: The formation of a shockwave requires energy, the temperature of the air increases sharply when it flows through a shockwave. The higher Mach number is, the stronger shockwave is, the higher temperature after the shockwave will be.

The formula of temperature change across a normal shockwave, Equation (7.36):

$$\frac{T_2}{T_1} = 1 + \frac{2(\gamma - 1)}{(\gamma + 1)^2} \frac{(\gamma M_1^2 + 1)}{M_1^2} (M_1^2 - 1)$$

shows that the temperature increase depends on the Mach number of the airflow. For example, if airflow's Mach number is 2, the air temperature will increase by 1.7 times. It means that at sea level, the temperature of 288 K (15°C) will increase to 486 K (213°C). If airflow's Mach number is 5, the air temperature will increase by 5.8 times: at sea level, the temperature will increase from 288 K (15°C) to 1670 K (1397°C).

The kinetic heating can be very intense at certain locations of a supersonic aircraft, for examples, nose of aircraft, leading edge of aerofoil. It can damage the materials and weaken the structure of the aircraft. Many methods to protect the supersonic aircraft from kinetic heating have been developed, and the following are some examples of the methods:

1) Materials: A material can become elastic and the strength of the material will be weakened before its temperature reaches its melting point. The higher its melting point is, the better it tolerates heat. Aluminum is a light metal, but its melting point is about 600°C. Titanium is light, and its melting point is over 1600°C, but it is very expensive. New materials for aircraft have been developed in recent years: alloys with titanium, which would be strong, flexible and can tolerate relatively high temperature. Composite with ceramic matrix is strong and can be used under high temperature environment, but it might not be flexible. Ultra-high-temperature ceramics (UHTCs) developed at NASA contain various metals can be used to protect the surface of high supersonic aircrafts.
2) Insulation: To cover the surface of the body parts, which experience the intensive kinetic heating, with the material, which can withstand high temperature with relatively low thermal conductivity, so heat will not penetrate to the body easily, and protect the flight instrument installed close to the "hot" wall of the body.
3) Surface radiation: To design the structure or shape of surfaces of the body, where are close to the heat sources, to radiate heat easily. The surface with a high emissivity can dissipates the heat quickly; or there is a large area of this surface structure, and the heat can be radiated out, and reduce the thermal load on the surfaces to prevent the local temperature from rapid increasing.

4) Surface cooling: To remove heat from the heated surface or structure of the body can prevent the rising of temperature. Behind the heated surface a jacket or a passage network can be installed, and coolant, which has got a relatively high heat capacity, or high latent heat, is driven through the passage/jacket. The coolant can absorb and transfer the heat from the surface to a heat sink to maintain the temperature in a safe level. For example, water can be used as a coolant, if the surface temperature is not too high.

Supersonic Control

Control of supersonic flight is a special branch of science. This chapter does not cover the principles of the stability and control for supersonic aircrafts. Only a few bullet points are mentioned here as a very general observation:

All-moving slab type of control surfaces: Each piece can be controlled individually.

Fully power-operated control surfaces: It would be quite heavy to move the control surfaces, when the aircraft flies in supersonic speed.

Synthetic stability and automatic control by on-board computer: The control require the precision to ensure the stability of the supersonic flight.

Appendix I: List of Derivatives

Differentiation (basic functions)

Function	*Derivative*
$f(x)$	$f'(x)$
C (C is a constant)	0
x^n (n is a constant)	nx^{n-1}
$\sin(x)$	$\cos(x)$
$\cos(x)$	$-\sin(x)$
$\tan(x)$	$\frac{1}{\cos^2 x}$
e^x ($\exp(x)$)	e^x
$\ln(x)$	$\frac{1}{x}$
$Af(x)$ (A is a constant)	$Af'(x)$
$Af(x)+Bg(x)$ (A and B are constants)	$Af'(x)+Bg'(x)$
Product rule: $f(x)g(x)$	$f'(x)g(x)+f(x)g'(x)$
Chain rule: $f(g(x))$	$f'(g(x))g'(x)$

Appendix II: θ–β–M Diagram

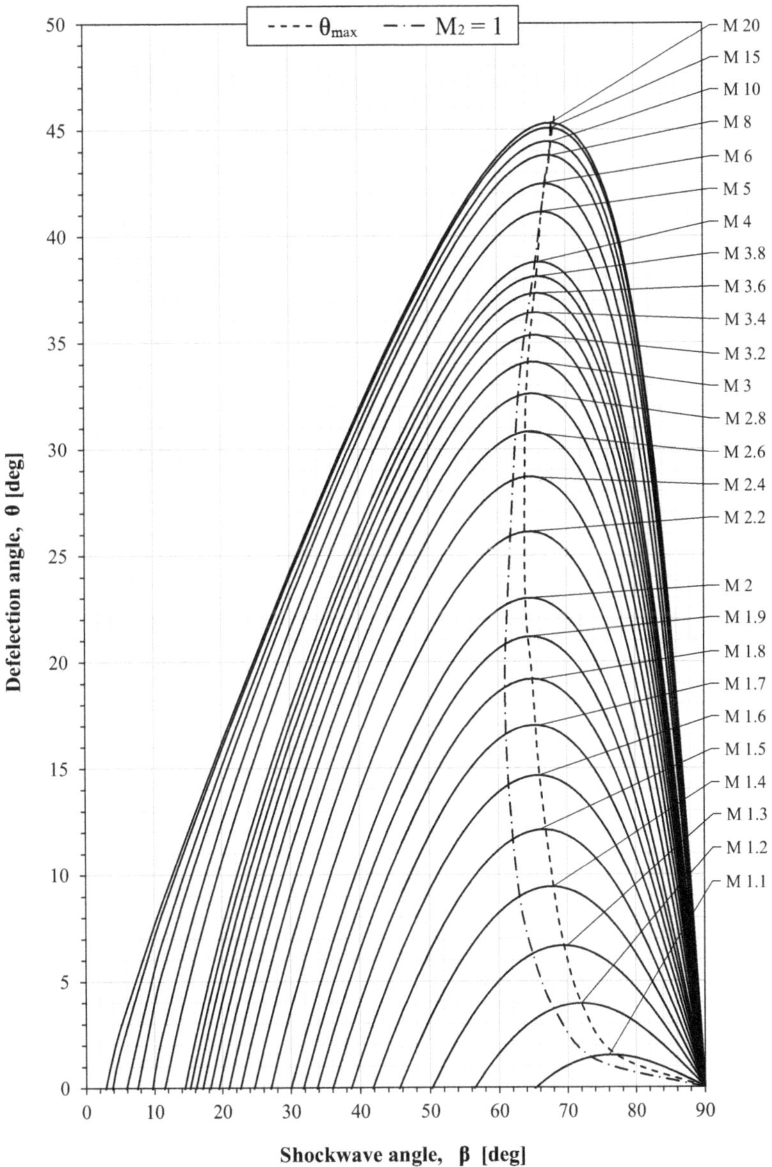

Appendix III: Prandtl–Meyer Function

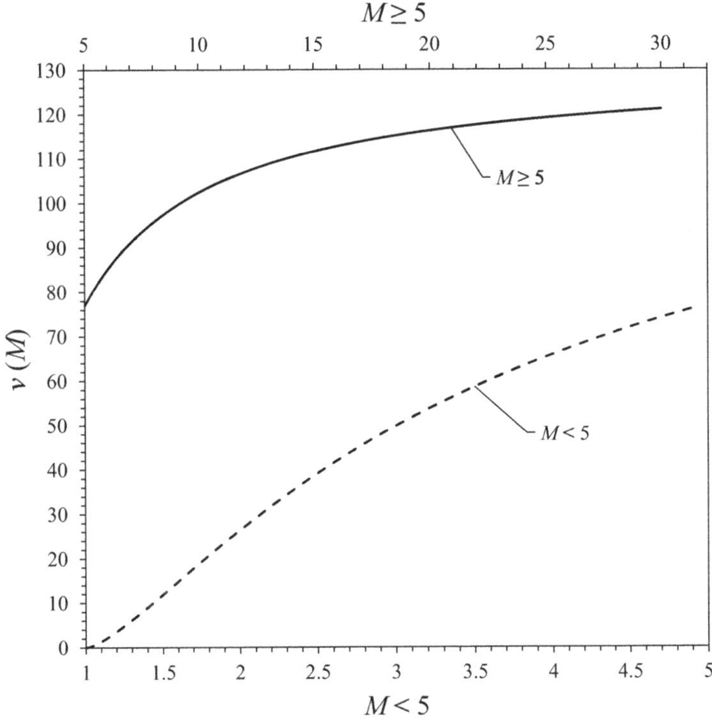

Appendix IV: Answers to Exercises

Chapter 1

1. $x=0, f(x)=x^2+1=1$, a local minimum.
2. $\frac{\partial f}{\partial x}=yz; \frac{\partial f}{\partial y}=xz; \frac{\partial f}{\partial z}=xy, df=yzdx+xzdy+xydz$.
3. $\int dp+\int vdv=p+\frac{1}{2}v^2+C$.

Chapter 2

1. The water speed inside the hose is 6.37 ms^{-1}, and the water speed at the outlet of the nozzle is 101.9 ms^{-1}.
2. $\rho_1=1.235$ kgm^{-3}, $T_1=290.7$ K (17.7°C), and $p_2=1.036\times10^5$ Pa.
3. The difference of static pressure/head measured by the venturi is 0.37 m (oil) (3088 Pa).
4. a. At the exit, airspeed is 150 ms^{-1} and air pressure is 8.97×10^4 Pa.
   b. Air temperature T_2 at the exit is 255.2 K.
   c. The airspeed is 108.83 ms^{-1}, and the density is 1.121 kgm^{-3}.
   d. The stagnation temperature is 289.8 K, and the stagnation pressure is 1.035×10^5 Pa.

Chapter 3

1. It is in transition at sea level ($\nu_{sea\text{-}level}=1.48\times10^{-5}$ m^2s^{-1}). At 10,000 ft, it is laminar at the same location with the same airspeed ($\nu_{10,000ft}=1.86\times10^{-5}$ m^2s^{-1}).
2. Not necessarily. The form drag of a smooth-surfaced round ball in an airflow can be relatively high.

3. a. Boundary layer leaves the surface.
   b. The rate of change of the speed in the direction perpendicular to the surface in a turbulent boundary layer is greater than that of in a laminar boundary layer, so it produces a higher skin drag (see Equation 2.1).
   c. Laminar boundary layer, if the speed of fluid is the same.
   d. The pressure distribution that produces lift is destroyed when the boundary layer separates over an aerofoil. There is a loss of lift, which leads to stall.

Chapter 4

1. NACA2412, because it has a greater thickness.
2. Two-dimensional airflow over a finite wing.
3. Because the pressure distribution is not symmetrical about its chord, the total lift is not "0" when AoA = 0°.
4. Induced drag is the horizontal (in the direction of RA) component of tilted lift caused by induced downwash (wingtip and trailing edge vortices).
5. Because induced drag is a component of lift.
6. A large *AR* makes the airflow over the aerofoil close to two-dimensional, and produces weaker vortices, than a weaker downwash. So it will increase the coefficient of lift and decrease the coefficient of induced drag.
7. Airspeed affects lift and drag, and the coefficient of total drag, but it does not affect the coefficient of lift.
8. AoA affects the coefficient of lift and the coefficient of total drag, so it affects both lift and drag.

Chapter 5

1. Centre of pressure changes with AoA, not with airspeed.
2. Fuselage produces aerodynamic force, which does not produce the restoring moment in rolling. Fuselage can be a stable feature to lateral stability, if the wing of an aircraft is at the upper part of the fuselage.

3. Yes, airspeed does effect pitch moment about AC.
4. The coefficient of pitch moment about AC is not considered to be affected by AoA.
5. Refer to the stable line in Figure 5.26.
6. Refer to the stable feature in Figure 5.21.
7. The AC is approximately located at 0.26 of chord = 0.468 m from the leading edge, and the CP at AoA = 4° is approximately at 0.39 of chord = 0.702 m from the leading edge.

Chapter 6

1. The airspeed is 292.57 ms^{-1}.
2. The true airspeed (TAS) is 150.76 ms^{-1} (293.3 kt).
3. $M_{fs} = 0.7$, $M_L = 1.06$, so this M_{fs} has exceeded M_{crit}.
4. The local speed of sound can be estimated by ISA if you know ALT. It is not always true that the speed of sound increases if both the air pressure and density are increasing.
5. Airspeed is decreasing while ascending; Airspeed is increasing while descending.

Chapter 7

1. It is approximately 277 K.
2. The speed coefficient $M^* = 0.73$.
3. a. $a_{throat} = 327\ ms^{-1}$;

   b. $\frac{A_{\text{inlet}}}{A_{\text{throat}}} \approx 1.06$;

   c. At the outlet of the nozzle: the airspeed is 403 ms^{-1}, and the air temperature is 239.2 K.
4. Approximately 0.74.
5. After the normal shockwave, the *Mach* number is 0.7, air pressure is 2.49×10^5 Pa, density is 2.28 kgm^{-3}, temperature is 380.2 K, and the airspeed is 274 ms^{-1}.

Chapter 8

1. 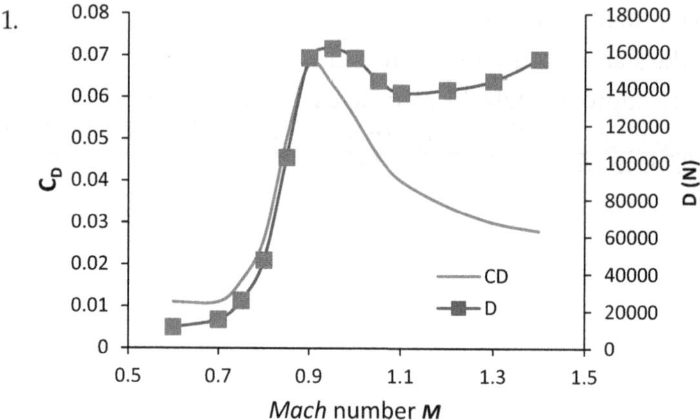

2. Refer to "Structure of Shockwave on Eerofoil" in Chapter 8.
3. Refer to "Changes of CP, C_L, and C_D between M_{crit} to M_{det}."
4. Wave dag and boundary layer separation drag caused by the two normal shockwaves formed on upper and lower surface of an aerofoil.
5. Bow shockwave is a combination of three pieces of shockwave: one normal shockwave in the middle (in front of a leading edge, or "nose"), and two oblique shockwaves at each side of the normal shockwave. The air temperature at the "nose" next to the leading edge is close to the stagnation temperature, which is higher that the free-stream air temperature, so the *Mach* number at the "nose" is lower than that in free-stream.
6. *Mach* tuck is the nose down pitch caused by normal shockwaves formed on an aircraft's wing. Adverse "stick force": the push force on the control column of an aircraft results in "pull" action, or vice versa, due to the shockwave formed on its elevator.
7. *Mach* buffet is the random vibration with loud noise an aircraft experiences, which is caused by the turbulent wake separation behind the normal shockwaves on aerofoil.

Chapter 9

1. The cross-over altitude of 225 kt/280 kt/M0.75 (approximately 30,000 ft) is higher than that of 250 kt/300 kt/M0.75 (approximately 27,000 ft).

2. At the altitude of Coffin Corner, an aircraft can only travel at one airspeed, and the aircraft will stall if it travels at a greater or smaller speed. At a cross-over altitude, an aircraft can travel with a wide speed range; only the method of monitoring the airspeed will change from this altitude (from kt to *Mach* number).
3. In short, a normal shockwave formed on a supercritical aerofoil is weaker than that on other aerofoil in transonic flight, so it causes less shock drag, i.e. less transonic drag.

Chapter 10

1. The stagnation pressure before the normal shockwave is 5.82×10^5 Pa, and the stagnation pressure after the normal shockwave is 4.73×10^5 Pa.
2. a. The deflection angle is 9°. The oblique shockwave is attached, because the maximum deflection angle at $M = 1.4$ is 9.5°, which is greater than the deflection angle 9°.
   b. The shock angle is approximately 63°.
   c. After the oblique shockwave, the air pressure is 1.668×10^5 Pa, temperature is 316 K, density is 1.84 kgm^{-3}, and the airspeed is 356.3 ms^{-1}.
3. a. The *Mach* number after the expansion fan is $M_2 = 2.1$.
   b.

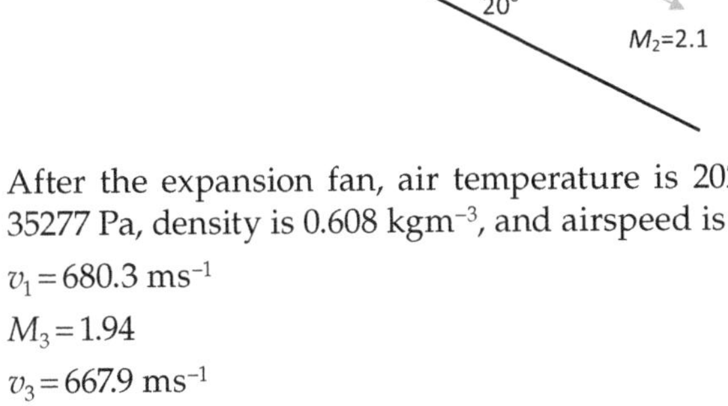

M_1=1.
45.6°
Expansion fan
28.4°
20°
M_2=2.1

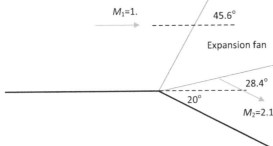

   c. After the expansion fan, air temperature is 202 K, pressure is 35277 Pa, density is 0.608 kgm^{-3}, and airspeed is 598.3 ms^{-1}.
4. a. $v_1 = 680.3$ ms^{-1}
   b. $M_3 = 1.94$
   c. $v_3 = 667.9$ ms^{-1}

References

AP3456, *Royal Air Force Manual*, The Training Group Defence Agency, UK, MOD Crown copy right, 1995.

Benenson, W., Harris, J. W., Stocker, H., and Lutz, H., editors, *Handbook of Physics*, Springer-Verlag, ISBN 0-387-95269-1, 2002.

Boys, S., and Wagtendonk, W., edited, *Principles of Flight and Aeroplane Performance*, PilotBooks, Vol. 7, 12th edition, Aviation Theory Centre (NZ) Ltd, ISBN 0-9583373-6-5, 2011.

Houghton, E.L., Carpenter, P.W., Collicott, Steven H., and Valentine, Daniel T., *Aerodynamics for Engineering Students*, 6th edition, Elsevier, 2012, = 19 in some places.

Jacobs, E.N., Ward, K.E., and Pinkerton, Robert M., *The characteristics of 78 related airfoil sections from tests in the variable-density wind tunnel*, NACA Report No. 460, NACA, 1933.

Kermode, A.C., *Mechanics of Flight*, 12th edition, Pearson, ISBN 978-0-273-77357-1, 2012.

Knight, K.C.D, Braun, K.M., Roy, Christopher J., Lu, Frank K., and Schetz, Joseph A., *Interference Drag Modelling and Experiments for a High Reynolds Number Transonic Wing*, 29th AIAA Applied Aerodynamics Conference, 27–30 June 2011, Honolulu, Hawaii.

Marzocca, P., *The NACA airfoil series*, Clarkson University, https://people.clarkson.edu/~pmarzocc/AE429/The%20NACA%20airfoil%20series.pdf, July 5, 2016.

Mayhew, Y.R., and Rogers, G.F.C., *Thermodynamic and Transport Properties of Fluids*, 2nd edition, Oxford, Basil Blackwell, 1968.

Panaras, Argyris G., *Aerodynamic Principles of Flight Vehicles*, American Institute of Aeronautics and Astronautics, Inc (AIAA), ISBN 978-1-60086-916-7, 2012.

Perry, J.H., *Chemical Engineers' Handbook*, 2nd edition, McGraw-Hill, 1941.

Prandtl, L., and Tietjens, O.G., *Applied Hydro- and Aeromechanics*, New York, Dover, 1957.

Robson, D., editor, *Aerodynamics, Engines and Airframe Systems for the Air Transport Pilot*, Aviation Theory Centre Pty Ltd, ISBN 1-875537-80-5, 2010.

Schlichting, H., *Boundary-Layer Theory*, 7th edition, McGraw-Hill, 1978.

Selig, M. S., Maughmer, M. D., and Somer, D. M., Natural-laminar-flow airfoil for general-aviation application, *Journal of Aircraft*, Vol. 32, No. 4, July–August 1995.

Serway, R.A., and Jewett, J.W., *Physics for Scientists and Engineers*, 7th edition, Thomson, 2008.

Shandong Engineering College and Shandong Electrical Power College, *Engineering Fluid-Dynamics*, Hydraulic Power Press, 1979 [in Chinese].

Strewart, J., *Calculus*, 6th edition, Thomson, 2009.

The Engineering ToolBox, http://www.engineeringtoolbox.com/sound-speed-solids-d_713.html, March 23, 2017.

Index

F

G

H

I

K

L

M

N

O

P

R

W

Y

Z

For Product Safety Concerns and Information please contact our EU representative GPSR@taylorandfrancis.com Taylor & Francis Verlag GmbH, Kaufingerstraße 24, 80331 München, Germany